Primate Archaeology and Adaptation

Primate Archaeology and Adaptation offers an interdisciplinary examination of Nicobar long-tailed macaques, situating their tool use, social behavior, and ecological flexibility within deep evolutionary time and contemporary conservation challenges. It documents the ecology, social organization, and behavioral repertoire of long-tailed macaques across Southeast Asia, including foraging strategies, food sharing, reproductive dynamics, exploratory behavior, and diverse forms of object manipulation. The book presents detailed accounts of tool-related behaviors, such as leaf 'glove' use, substrate selection for pounding, and the innovative use of anthropogenic materials, while carefully distinguishing between tool use and broader object use. By bridging fossil history, behavioral ecology, and archaeological method, it rethinks material culture beyond humans.

This book will interest scholars and advanced students in biological anthropology, primatology, archaeology, human evolution, animal behavior, and conservation studies. It will also appeal to researchers working on material culture, island ecology, and South Asian environmental policy.

Jayashree Mazumder is an assistant professor in the Department of Anthropology at the University of Delhi. She was formerly an ANRF/SERB National Postdoctoral Fellow at the Centre for Ecological Sciences, Indian Institute of Science (IISc), Bengaluru and completed her PhD from Indian Institute of Science Education and Research (IISER), Mohali. A biological anthropologist and primatologist, her research spans primate archaeology, tool use, and material culture in Nicobar long-tailed macaques, and currently centers on the movement ecology and conservation of hoolock gibbons in Northeast India.

Primate Archaeology and Adaptation

A Study of Nicobar Long-Tailed Macaques in India

Jayashree Mazumder

Routledge
Taylor & Francis Group
LONDON AND NEW YORK

First published 2027
by Routledge
605 Third Avenue, New York, NY 10158

and by Routledge
4 Park Square, Milton Park, Abingdon, Oxon, OX14 4RN

Routledge is an imprint of the Taylor & Francis Group, an informa business

ISBN: 9781041014010 (hbk)
ISBN: 9781041014027 (pbk)
ISBN: 9781003614579 (ebk)

DOI: 10.4324/9781003614579

Typeset in Sabon
by Deanta Global Publishing Services, Chennai, India

Dedicated to Dr. Jane Goodall, Dr. P. K. Seth, and Dr. A. P. J. Abdul Kalam – visionaries who embodied compassion, scientific curiosity, and a commitment to understanding life in all its forms. Their work continues to inspire the pursuit of knowledge that bridges behavior, science, and humanity.

Figure 0.1 Portrait of a Young Nicobar Long-Tailed Macaque (*Macaca fascicularis umbrosus*) in a Moment of Stillness. The Image Captures the Species' Characteristic Facial Features and Alert Expression, Offering a Glimpse into the Complex Social and Cognitive Life of This Adaptable Primate.

Contents

Figures

Tables

Acknowledgments

This monograph would be incomplete without the support, guidance, and encouragement of the many individuals I have met over the past nine years. Since 2017, people from diverse walks of life have contributed to this work in various ways, and I take this opportunity to express my sincere appreciation.

I am deeply grateful to my supervisor, Dr. Parth R. Chauhan, for taking me under his mentorship despite our differing research interests. His guidance shaped the core of this thesis – from teaching me how to think critically and work independently to encouraging me to constantly strive for better. Though primatology is outside his primary area, his incisive questions and unwavering support pushed the boundaries of my thinking. More than a mentor, he has been a friend and a source of strength throughout this journey, and I remain sincerely grateful to him.

My heartfelt thanks to Dr. Stefano S. K. Kaburu (Nottingham Trent University, UK) for his mentorship, insightful feedback, and constant encouragement throughout the PhD. I also extend my sincere gratitude to the Department of Humanities and Social Sciences at IISER Mohali and to my research committee members – Dr. Manjari Jain, Dr. Ritajyoti Bandyopadhyay, and Dr. Debdulal Saha – for their guidance, critique, and support.

Fieldwork in the Andaman & Nicobar Islands brought me into contact with many individuals who made this research possible. I am grateful to the forest departments for research permissions, and especially to Mr. Vinod and Mrs. Dakshmi (Nicobar) for their help in the field. Special thanks to Amit Kerketta, Uday Mondal, Mukesh, and the families of Mr. Papa Rao and Mr. Srinu for their hospitality. I remain in deep gratitude to the tribal chief and the villagers of Makacua (Little Nicobar) who welcomed me as one of their own. I also thank Mr. Sameer and Mr. Rohit for their local support.

I am immensely thankful to my field team – Nirmal Nath Upadhyay (Research Assistant) and interns (Kiran and Diksha Mandik) – for their dedication, adaptability, and friendship. This work would not have been possible without their cooperation, particularly in such remote and challenging field sites. At Port Blair, I extend thanks to Pondicherry University for hosting me, and to students Sayeri Paul, Shreeja Rini, Shrusti Birari, Samrat, Debarati Paul, and Bidisha Sinha for their camaraderie.

I would also like to thank my friends and colleagues – Dr. Tritisri Das, Sanchari Bhattacharyya, Dr. Shilpi Rajpal, Trirupa Chakraborty, Martina Narzary, Angelina Lakra, Ashley Correia, Dr. Subhashri Sarkar, Dr. Ankur Parashar, Dr. Yezad Pardiwalla, and Dr. Rajesh Poojari – for being my anchor during difficult times.

My heartfelt gratitude goes to my mother, whose quiet strength and life lessons shaped how I navigated this journey, and to my late father, whose teachings remain with me even in his absence. I thank my entire family for believing in me.

I gratefully acknowledge the funding agencies: IISER Mohali, the International Primatological Society (UK), the Leakey Foundation (USA), USIEF (Fulbright-Nehru Visiting Scholar Program), and IUCN SSA. Special thanks to Pratibha Nair (USIEF), Dr. Sudarsan Dash (USIEF), and Dr. Susan Cheyne for their encouragement and support.

Lastly, I thank the primates – the long-tailed macaques – who unknowingly shared their lives, tolerated my presence, and made this research possible. To all those mentioned here, and those I may have unintentionally left out, I extend my deepest gratitude.

Jayashree Mazumder

1 Primate Archaeology across India and Abroad

Primate archaeology is an emerging field that explores the material culture and behavioral evolution of non-human primates, offering crucial insights into the cognitive and cultural development of our closest animal relatives (Carvalho & Almeida-Warren, 2019; Haslam et al., 2009, 2017; Pascual-Garrido et al., 2023). In India, the study of primate archaeology has gained increasing significance due to the country's rich primate biodiversity, which includes species such as the Nicobar long-tailed macaques (*Macaca fascicularis umbrosus*), bonnet macaques (*Macaca radiata*), and lion-tailed macaques (*Macaca silenus*). These species have demonstrated remarkable behaviors, such as tool use and object manipulation, which offer parallels to early human technological evolution.

Primate archaeology not only reveals how non-human primates have adapted to environmental challenges over millennia but also highlights the role of cultural transmission in shaping these behaviors (Haslam, 2024). The archaeological record of tool use – such as stones used for cracking open hard-shelled food – provides a long-term view of how primates have developed and passed down these skills through generations (Luncz et al., 2024). By examining both fossil evidence and contemporary tool-use sites, researchers can trace the potential origins of these behaviors and their connection to survival and adaptation (Proffitt et al., 2022).

This chapter discusses the key paleoanthropological discoveries in India, methodologies, and evolutionary implications of primate archaeology by positioning it within the broader discourse on cognitive evolution and material culture in non-human primates. Through this approach, we can better understand the evolutionary roots of human-like behaviors and the profound connections between culture, cognition, and adaptation in the animal kingdom.

DOI: 10.4324/9781003614579-1

1.1 The Fossils and the Living

The study of Asian anthropoids occupies a pivotal position in the field of anthropology, especially for those examining the evolutionary history and diversity of primates in this region. This pursuit significantly enhances academic discussions about primate biology, adaptation, and evolution, integrating diverse disciplines such as biology, paleontology, and genetics. These interdisciplinary approaches not only shed light on the past but also illuminate the present and project into the future biodiversity of Asia. Jean-Jacques Jaeger, a luminary in the field of paleontology, opens our eyes to this lost world. Jaeger and colleagues propose that some of the earliest anthropoid fossils in Asia (e.g., *Bahinia*, *Eosimias*, *Afrasia*) suggest that the group ancestral to monkeys, apes, and humans may have originated in Asia and later radiated into Africa – implying that ancient Asian anthropoids might have played a pivotal role in shaping the evolution of modern higher primates (Chaimanee et al., 2012). Yet, astonishingly, the story of these creatures, once flourishing in areas now known as India, Pakistan, Sri Lanka, and the broader South Asian region, remains largely untold due to their neglect in scientific exploration.

India, in particular, has emerged as a pivotal region for primate fossil discoveries, providing invaluable insights into the evolution and diversity of primates in South Asia. Imagine stepping back in time, millions of years ago, to a world where the landscapes of Asia were teeming with a rich diversity of primate life, much more varied than we could have imagined. Now, let us explore the tales of two fascinating characters from this prehistoric drama: *Ramapithecus* (discovered in 1932) and *Sivapithecus* (discovered in 1982) (Bhandari et al., 2018; Gingerich, 1984). Imagine finding fragments of a jaw that once belonged to *Ramapithecus* (now taxonomically merged with *Sivapithecus*) in the Siwalik hills of northern India. Initially mistaken as a progenitor to humans, this fossil primate dating to the Middle and Late Miocene epochs was later linked closer to the lineage leading to orangutans (Almécija et al., 2021; Begun, 2015; Marzke, 2013). The plot thickens when Elwyn Simons, an American anthropologist with an eye for evolutionary puzzles, posits that *Ramapithecus* might represent the earliest fork in the road, leading away from the lineage that would eventually evolve into modern humans and apes (Kay, 2018; Kennedy & Ciochon, 1999).

Not far from the story of *Ramapithecus* is that of *Sivapithecus*, another fossil primate whose remains sprinkle across the Siwalik Hills of northern Pakistan, Gujarat and even further afield in Turkey, China, and Kenya. Close enough to *Ramapithecus* to be considered possibly the same species by some but distinct enough to others to represent a direct ancestor of today's orangutans, *Sivapithecus* adds another layer of intrigue to our understanding of primate evolution. By the 1980s, evolving interpretations and fresh discoveries painted *Sivapithecus* firmly as an ancestor of the modern orangutan, with roots tracing back around 17 million years. *Sivapithecus* fossils offer essential insights into the evolutionary history of great apes. Although not a direct ancestor of humans, *Gigantopithecus* (discovered in 1934), another notable fossil, is considered one of the largest known primates to have ever lived. *Gigantopithecus* fossils, dating back to the Miocene and Pleistocene epochs, have been discovered in China and India (Pilgrim, 1915; Zhang et al., 2024). In India, they became extinct in the Miocene, while they lasted until the Late Pleistocene and were contemporaneous with hominins in China. These immense apes likely had a herbivorous diet and thrived in forested environments. Initially referred to as *Indopithecus*, these genera represent the early stages of ape evolution. The existence of *Kenyapithecus* (discovered in 1961), which shares similarities with *Indopithecus* and has been found in parts of India and Africa, suggests connections between primate populations across continents.

Advancing the timeline to 2014, a partial right mandible discovery in India introduced a new genus and species near the base of the *Sivaladapinae* subfamily, offering fresh perspectives on ancient primate dynamics in the region (Roberts et al., 2014). Even more compelling is the scene uncovered at the Vastan lignite mine in Gujarat: the oldest known euprimates from India, dating back some 53 million years, offer a window into life just as the Indian plate collided with Asia (Bhandari et al., 2018; Gilbert et al., 2017). *Marcgodinotius indicus* and *Asiadapis cambayensis*, among these primordial finds, challenge our assumptions about the origins and journeys of these primates. A complex story of migration and evolution unfolds, stretching across continents from Europe to Asia, and spanning epochs from the Paleocene to the Eocene. In the Pleistocene epoch, ancient members of the *Homo* genus, including *Homo erectus*, populated regions of India. Discoveries from fossil sites like the Narmada Valley and the Siwalik Hills have shed light on the migration and adjustment of early humans in South Asia (Lewis, 1937; Sonakia & De Lumley, 2006).

Contemporaneous with early hominins in the Pleistocene Siwaliks was *Theropithecus* sp., a large-bodied papionin closely related to the modern gelada of Ethiopia. Fossil mandibles and dental remains recovered from the Siwalik Hills (Potwar Plateau, Pakistan, and northern India) reveal the characteristic high-crowned molars of this genus, an adaptation to grazing on tough C4 grasses (Delson, 1993). Stable isotope analyses further confirm its specialized grassland diet, making it one of the few primates so strongly tied to open savanna ecosystems (Patnaik et al., 2014). The eventual extinction of *Theropithecus* outside Africa underscores the ecological vulnerability of hyper-specialized grazers during periods of climatic and environmental change (Elton, 2002; Getahun et al., 2023; Hughes et al., 2008).

Most recently, a new fossil primate named *Kapi ramnagarensis* was discovered in 2015 in northern India (Gilbert et al., 2020, 2025). It is estimated to have lived around 12.5–13.8 million years ago during the Middle Miocene epoch. This primate is distinct from all other known fossil apes. Its complete lower molar was unearthed in 2015 from the Lower Siwaliks of Ramnagar in Jammu and Kashmir, India. This discovery marks the first new hominoid species found at the Ramnagar site in nearly a century and the first new Siwalik ape species in over three decades.

The discovery of each fossil, each fragment of jaw or tooth, serves as a gateway to an ancient world, providing us with a glimpse of the remarkable diversity and evolutionary tales that have influenced the primate lineage (Mercader et al., 2007). The recent findings in Asia represent the beginning of a narrative that is still unfolding, encouraging us to explore, question, and admire the intricate tapestry of life that has influenced our understanding of the world. These fossil discoveries from India play a significant role in enhancing our comprehension of primate evolution, including the emergence of early hominins and their spread across Asia.

Continuing research and new discoveries are persistently refining our understanding of India's contribution to the evolutionary history of both non-human primates and human primates. The transition from fossil primates to living primates in India offers a bridge between past and present, connecting ancient evolutionary lineages to the modern primate species that inhabit the region today. Fossils serve as the critical link in this narrative, allowing us to trace the evolutionary path that has shaped the diversity of primates we see in contemporary India. As we move from studying ancient, fossilized remains to observing living species, we uncover not just the biological traits passed down through generations but also the behavioral and ecological adaptations that have allowed primates to thrive in various environments.

India's primate fossil record, rich with species like *Sivapithecus*, reveals a time when the subcontinent was home to primates that bore close affinities to the ancestors of both modern apes and early humans. These fossils not only inform us about the physical characteristics of these primates but also offer clues about their diets, habitats, and social structures (Patnaik et al., 2014). By comparing fossilized remains with living species such as macaques, langurs, and gibbons, we can piece together how these early primates adapted to changing environments, evolving traits that would later become critical for survival in the present day.

This continuity between ancient and living primates is not just biological but also behavioral. Living primates in India, such as the lion-tailed macaque, Nicobar long-tailed macaque, and bonnet macaque, exhibit complex behaviors such as tool use, social learning, and intricate foraging strategies – behaviors that may have deep evolutionary roots (Chatterjee,

2003, 2012; Dileep & Jose, 2014; Hohmann, 1988; Mazumder & Kaburu, 2020; Sinha, 1997). The archaeological record, combined with modern observations of primate behavior, suggests that some of these practices may have been shared by their ancient ancestors (Luncz et al., 2015). Fossils, alongside living primates, help us understand the cognitive and social evolution that shaped the primate lineage (Carvalho & Almeida-Warren, 2019). The study of primate archaeology has shown that primates possess the ability to adapt culturally and cognitively to their surroundings, a trait that was likely crucial for early hominins as well (Haslam et al., 2009).

While fossil discoveries allow us to reconstruct the biological lineage of primates, studying living primates enables us to observe evolutionary processes in action. The behaviors seen in contemporary primates may be reflections of adaptations developed over millions of years in response to environmental pressures. For instance, the use of tools by Nicobar macaques to crack open hard-shelled food items echoes the foraging strategies that might have helped their ancient ancestors survive periods of scarcity or environmental change (Das et al., 2020). This behavioral continuity highlights the dynamic interplay between ecology, cognition, and survival that has been present throughout primate evolution.

Evolutionary anthropological studies (combining both fossil primates and living primates) in India illuminate the continuity of life and evolution in the region. Fossils provide the foundational narrative, offering insights into the physical and ecological adaptations of ancient primates while living species carry forward these adaptations, evolving new behaviors to meet the challenges of their environments. Together, they create a rich tapestry of evolutionary history that connects the deep past to the vibrant ecosystems of modern India and sets the stage for an in-depth exploration of early tool use and the development of material culture in non-human primates. This chapter explores these modern findings and their implications for the broader understanding of primate cognition, material culture, and adaptive strategies.

1.2 Known Tool Use and Material Culture in Indian Primates and Beyond

The study of tool use in non-human primates has long been a window into understanding cognitive evolution, material culture, and adaptive behaviors in the animal kingdom (Shumaker et al., 2011). In this respect, primate archaeology is an emerging field that intersects anthropology, archaeology, and primatology to explore the evolutionary history of non-human primates through their material culture and behavioral evidence (Haslam et al., 2017). Globally, this discipline has challenged traditional boundaries by demonstrating that tool use and other cultural behaviors are not exclusive to humans.

By studying non-human primates such as chimpanzees, orangutans, capuchins, and macaques, researchers have uncovered long-term traditions of tool use and ecological adaptations that mirror early human behaviors (Albanese et al., 2022; Babitz, 1999; Boesch, 2012, 2017; Call & Tomasello, 1994; Elisabetta et al., 2013; Leca et al., 2021; Laura et al., 2026).[*] Fossil records, ancient tools, and ecological data provide a unique lens to trace the cognitive and social evolution of these species, offering insights into how non-human primates interact with their environments, develop culture, and adapt to changing landscapes. This comparative approach enriches our understanding of the origins of human technology and social complexity by placing them within a broader evolutionary framework. As the field continues to expand, primate archaeology provides crucial data for reconstructing not only the behavior of our closest evolutionary relatives but also the broader context of cultural and technological evolution across species. In India, the study of tool use in non-human primates extends beyond the Nicobar long-tailed macaques and includes significant observations from other species like the bonnet macaques and lion-tailed macaques. In one of the earliest documented cases of tool use among bonnet macaques, the author reported a wild individual manufacturing a rudimentary tool to access food, a behavior that emphasizes their adaptive abilities and adaptive strategies (Sinha, 1997). Similarly, in 1988, a study on lion-tailed macaques revealed their use of simple tools, such as sticks, to access insects hidden within crevices (Hohmann, 1988). These observations broaden the understanding of material culture and cognitive abilities in non-human primates in India.

Both examples highlight that tool use is not isolated to one species but spans across different primate taxa in the region. These findings also underscore the importance of ecological pressures in shaping tool-using behavior, as bonnet macaques and lion-tailed macaques inhabit diverse environments, from urbanized landscapes to dense forests. Together with research on the Nicobar long-tailed macaques, these studies provide a comprehensive view of the early stages of material culture in Indian primates, enriching the broader discourse on primate archaeology and behavioral evolution.

Within the context of India, the unique behavior of macaques, especially the Nicobar long-tailed macaques, offers an unparalleled opportunity to explore the evolution of early tool use in primates. These macaques, native to the Nicobar Islands, exhibit sophisticated behaviors related to object manipulation and tool use, particularly in accessing food resources (Das et al., 2020; Mazumder & Kaburu, 2020; Pal et al., 2017). Such behaviors are not only observed in the wild but are also recorded archaeologically through wear patterns on stones, bones, and other materials (Boesch & Boesch, 1990; Lemorini et al., 2014). The discovery of long-tailed macaque tool-use sites in Thailand has spurred an interdisciplinary approach that integrates behavioral ecology with archaeological techniques to uncover the cognitive and cultural evolution of these primates (Haslam et al., 2013). Building on these findings, the Nicobar long-tailed macaques, a geographically and behaviorally distinct subspecies, offer a unique case study due to their isolation in the Nicobar Islands. Long-tailed macaques are notable for their frequent use of tools in everyday foraging activities, particularly their use of stones and other raw materials to crack open hard-shelled foods such as nuts, mollusks, and crabs (Gumert & Malaivijitnond, 2013; Tan et al., 2015; Velankar et al., 2023). Such behavior demonstrates a remarkable level of adaptive strategy, showcasing the macaques' ability to manipulate their environment to access difficult-to-obtain food resources. These activities involve selecting appropriate tools – typically stones of a certain size and weight – and using them repeatedly for specific tasks, illustrating a form of material culture transmitted across generations.

Field studies done in Southeast Asia have documented these tool-use behaviors over long periods, revealing that individual macaques exhibit preferences for particular tools and methods of cracking or extracting food. For example, some macaques consistently select stones of a particular texture or weight, which is believed to be related to their efficiency in breaking shells without damaging the food inside (Gumert & Malaivijitnond, 2013; Malaivijitnond et al., 2007; Tan et al., 2015; Velankar et al., 2023). Additionally, tool use is not limited to a single season or food resource; it is observed year-round, indicating that this behavior is an embedded part of the macaques' ecological adaptations rather than a sporadic or opportunistic occurrence.

1.3 The Archaeological Record of Tool Use: Methods in Primate Archaeology of Long-Tailed Macaques

The intersection of archaeology and behavioral ecology is critical for understanding the evolution of tool use in primates (Urbani et al., 2022; Whiten, 2013). Indeed, archaeological evidence extends from the earliest known stone tools, dated to 3.3 million years ago at Lomekwi in Kenya (Harmand et al., 2015), to the oldest archaeological record of non-human primate tool use at chimpanzee nut-cracking sites in West Africa, dated to approximately 4,300–4,500 years ago (Mercader et al., 2007). More recently, wild capuchin monkeys in Brazil have been documented unintentionally producing sharp stone flakes – 'pseudo-tools' – as byproducts of percussive behaviors while accessing mineral resources (Proffitt et al., 2016). Archaeological methods, such as the analysis of wear patterns on tools, offer insights into how and why certain behaviors evolved. In the case of the long-tailed macaques, these methods provide a historical context for the observed behaviors, allowing researchers to track changes in tool use over time and across different environments.

For instance, studies of tool-use sites in the Nicobar Islands have revealed that macaques preferentially select tools based on the availability of raw materials (Mazumder & Kaburu, 2021). In areas where suitable raw materials are abundant, macaques exhibit more complex tool-use behaviors, while in regions where such materials are scarce, they rely more on opportunistic foraging strategies (Mazumder & Kaburu, 2021). This variation highlights the adaptability of the species and suggests that environmental pressures play a significant role in shaping material culture.

Moreover, the archaeological record can help identify periods of innovation or cultural change within macaque populations. As primates are subject to ecological and social pressures, shifts in tool use may reflect broader changes in group dynamics, resource availability, or environmental stressors (Arre et al., 2021; Matsumura, 1999). By combining behavioral observations with archaeological data, researchers can construct a more comprehensive picture of how tool use functions within the broader ecological context of the Nicobar Islands.

While behavioral observations provide direct evidence of tool use, archaeological studies offer a long-term perspective on how these behaviors have evolved and persisted. In Thailand, for example, researchers have discovered tool-use sites where macaques repeatedly return to access food resources (Luncz et al., 2017, 2019). These sites often contain clusters of stone tools that show wear patterns consistent with long-term use for cracking shells or breaking apart other food items (Haslam et al., 2013). By analyzing these stones, researchers can trace the history of tool use in specific macaque populations and draw parallels to early human tool use.

The wear patterns on the stones reveal significant information about the types of forces exerted on the tools and the nature of the objects being processed. For instance, stones used to crack nuts show distinctive impact marks and flaking, while those used for smashing open crabs or mollusks exhibit different forms of damage (Boesch & Boesch, 1990; Haslam et al., 2013; Lemorini et al., 2014). Stones used by non-human primates for cracking open food items exhibit distinctive wear patterns that can be analyzed to infer their function (Dubreuil & Nadel, 2015). Likewise, tools used to crack nuts typically show impact fractures, flaking, and pitting on their surfaces, while those used to open crabs or mollusks may exhibit different types of damage, such as edge wear or abrasions (Arroyo et al., 2016; Bril et al., 2012; Tan et al., 2015). These marks are analyzed under high magnification to determine the exact forces applied and the types of materials processed. These variabilities in tool wear highlight the primates' ability to modify their tool use based on the task at hand (Bjorklund & Gardiner, 2010; Diogo et al., 2012), demonstrating a form of behavioral flexibility. Importantly, the discovery of such sites also suggests that these tool-using behaviors have been practiced for several generations, possibly indicating a form of cultural transmission where young macaques learn by observing and mimicking older individuals (Biro et al., 2003; Gruber et al., 2011; Gumert & Malaivijitnond, 2012; Haslam et al., 2013; Luncz et al., 2017).

In the Nicobar Islands, certain areas have been identified as key tool-use sites (Mazumder, 2022; Mazumder & Kaburu, 2020, 2021), where macaques repeatedly gather to forage and use tools to process food. These sites often include clusters of tools that display wear patterns indicative of prolonged use, and their spatial distribution suggests that these areas serve as 'processing centers' for food extraction. Unlike random occurrences of tool use, these sites are characterized by a high concentration of artifacts, which are consistently used by different macaques over time.

To investigate these sites archaeologically, researchers employ systematic excavation techniques, similar to those used in human archaeology (Mercader, 2002; Mercader et al., 2007). These methods involve careful mapping and recording of the location of tools, as well as the stratigraphy and sedimentology of the area. By excavating these sites, archaeologists can determine how long tool use has been occurring, whether the macaques are selecting specific locations for tool use, and whether there is evidence of site reuse over generations (Haslam et al., 2013; Proffitt et al., 2024; Reeves et al., 2024). One of the most critical aspects of studying tool use in Nicobar long-tailed macaques is the analysis of the tools themselves. Artifact analysis focuses on examining the physical characteristics of the tool, including its size, shape, and material composition (Luncz et al., 2022). In primate archaeology, this analysis is combined with wear pattern studies to assess how the tools were used and how long they have been in use (Haslam et al., 2013; Reeves et al., 2024).

In the case of long-tailed macaques, tools are often selected based on their size and weight, and researchers have observed that some tools show more advanced forms of wear, suggesting prolonged and repeated use (Gumert & Malaivijitnond, 2013). This wear pattern analysis provides insight into the cognitive abilities of the macaques, indicating their ability to choose the most effective tools for specific tasks and their capacity to maintain and reuse tools over time.

The spatial distribution of tools at Nicobar macaque sites is another critical aspect of archaeological research. By mapping the locations of tool-use areas, researchers can investigate whether macaques exhibit preferences for certain environments or resources. For example, many of the tool-use sites are located near water sources, where mollusks and crabs are abundant and also face high rates of anthropogenic pressure (Mazumder, 2022; Mazumder & Kaburu, 2020). These locations are revisited regularly by macaques, suggesting that they recognize specific areas as productive food sources.

The study of spatial distribution also sheds light on whether tool use is a culturally transmitted behavior. If macaques consistently use the same locations and tools across multiple generations, it indicates social learning and site fidelity, reflecting a form of cultural transmission (Luncz et al., 2019; Reeves et al., 2023; van Schaik & Pradhan, 2003). Archaeologists record the positioning of tools in relation to the surrounding environment, noting recurring patterns in tool selection, clustering, and site reuse (Luncz et al., 2019; Reeves et al., 2023).

1.4 Cultural Transmission and Material Culture in Non-Human Primates

One of the most compelling aspects of tool use in non-human primates is the potential for cultural transmission. Cultural transmission refers to the passing of behaviors and knowledge from one generation to the next, often through social learning (Aplin, 2016). In humans, material culture – tools, artifacts, and other physical manifestations of culture – is a key indicator of this process. Although non-human primates do not produce tools as sophisticated as those of early humans, their consistent and repetitive use of objects like stones points to a rudimentary form of material culture (Bandini et al., 2022; Luncz et al., 2015). In this book, culture in non-human primates is treated conservatively, referring to population-level behavioral variation that is socially influenced and persists beyond individual lifespans, without presuming intentional teaching or human-like symbolic systems.

In the case of non-human primates, young individuals are often seen observing adults during foraging activities, sometimes mimicking their use of tools to access food (Agostini & Visalberghi, 2005; Bandini & Tennie, 2020; Galef & Giraldeau, 2001; O'Malley & Fedigan, 2005; Rapaport & Brown, 2008). This observational learning is critical to the development of tool-use skills and suggests that these behaviors are not entirely innate but are instead learned through social interaction. The transfer of tool-use knowledge within primate groups implies that these behaviors are culturally inherited, making them an essential component of the species' adaptive strategy.

Comparisons to other tool-using primates, such as chimpanzees and capuchin monkeys, further underscore the significance of this behavior in macaques. In each case, tool use is closely linked to environmental challenges and the availability of food resources. However, the persistence of these behaviors across generations highlights the role of culture in shaping primate societies. Just as early humans used tools to enhance their survival and adapt to new environments, the Nicobar long-tailed macaques use tools to exploit niche food sources that would otherwise be inaccessible, suggesting a parallel form of material culture in these non-human primates.

1.5 Linking Past and Present: Conservation Implications

Conservation in Southeast Asia has reached a state of urgency due to rapid deforestation, habitat fragmentation, and illegal wildlife trade, all of which threaten the region's rich biodiversity (Gomez et al., 2023; Hiller et al., 2004; OECD, 2019; Tacconi, 2007). This area, home to some of the world's most endangered species, including many primates, is losing its forests at an alarming rate, pushing species like gibbons, orangutans, and macaques closer to extinction. The degradation of these ecosystems not only impacts the species that inhabit them but also disrupts ecological processes like seed dispersal, which are crucial for maintaining forest health. Within this context, primate archaeology offers critical insights into how primates have adapted to and interacted with these environments over time (Behie et al., 2019; Bezanson & McNamara, 2019). By studying the material culture and long-standing behavioral traditions of primates in Southeast Asia, such as tool use in macaques, researchers can better understand the ecological roles these primates play and how their loss could impact ecosystem stability. Conservation efforts, informed by findings from primate archaeology, can help protect not only the species but also the cultural and ecological heritage of these regions, making the preservation of both biological diversity and behavioral history an intertwined priority.

Primate archaeology, particularly in the context of the Nicobar long-tailed macaques, faces several challenges. The remote and fragile ecosystem of the Nicobar Islands limits the accessibility of research sites, and the tropical climate can accelerate the degradation of organic materials, making long-term preservation difficult. Moreover, distinguishing between natural wear and tool-use wear on stones and other perishable raw materials such as wood can be challenging, requiring careful analysis and corroboration with behavioral studies. One of the key challenges is the government-imposed restricted access to the Nicobar Islands, aimed at safeguarding the indigenous communities and preserving the delicate ecosystems. This restriction prohibits any form of excavation and invasive study on the animals. Additionally, the removal of artifacts from their original context is strictly forbidden, ensuring the protection of both the cultural heritage and the environmental integrity of the region.

Despite these challenges, the field of primate archaeology holds significant promise for expanding our understanding of non-human primate cognition and material culture. Understanding the evolutionary behaviors of India's primates through archaeological evidence not only deepens our comprehension of primate cognition and material culture but also holds profound implications for conservation efforts. In particular, the archaeological record of tool-using primates, such as the Nicobar long-tailed macaques, provides a valuable foundation for conservation strategies aimed at preserving both species and their unique cultural behaviors. By studying the behavioral evolution of primates in relation to ecological changes over millennia, we can better address the challenges these species face in modern ecosystems, where habitat destruction, climate change, and human-wildlife conflict increasingly threaten their survival. Importantly, documenting tool use and material traces in primates strengthens conservation arguments, not by romanticizing behavior, but by identifying specific sites, resources, and ecological contexts that sustain these behaviors. Such knowledge allows conservation planning to move beyond species-level protection toward safeguarding behavioral diversity and the habitats that enable it.

1.5.1 The Importance of Protecting Tool-Using Species and Cultural Traditions

Tool-using primates, such as long-tailed macaques (Figure 1.1), chimpanzees, capuchins, etc., are not only biologically significant but also represent unique examples of non-human cultural traditions (Whiten, 2013). The use of tools to crack open nuts, mollusks, and crabs is a behavior that is passed down across generations, indicating the presence of social learning and cultural transmission within macaque populations (Luncz et al., 2017). Protecting these macaques is, therefore, not just a matter of preserving a geographically isolated species, but also of safeguarding their behavioral and adaptive traditions, which could be easily lost if these populations dwindle.

This highlights the need to conserve primate habitats, especially those that support tool-using populations. The Nicobar Islands, home to these macaques, are ecologically fragile and susceptible to human encroachment, development projects, and natural disasters (Bridger, 2022; Murty et al., 2006; Sivakumar, 2010; Velankar et al., 2016). Habitat destruction can lead to the loss of critical resources, such as the specific stones and food sources needed for macaques to continue their tool-using behaviors (Gumert et al., 2013; Luncz et al., 2017). In preserving the macaques' habitat, conservationists ensure that the conditions necessary for the persistence of this unique cultural trait are maintained. Additionally, the

Figure 1.1 A Silhouetted Study Subject Pauses on a Bamboo Scaffold at Dusk, Its Quiet Form Etched against the Forest Backdrop.

protection of tool-using primates offers insights into broader conservation priorities. The behaviors of these primates serve as indicators of the health of their ecosystems. A decline in tool-using behaviors could signify disruptions in resource availability, environmental stress, or habitat fragmentation, issues that are crucial for conservationists to address in order to maintain biodiversity and ecosystem balance (Luncz et al., 2017, 2019).

1.5.2 Insights from the Archaeological Record: Learning from the Past

The archaeological record of India's primates provides invaluable data for conservation efforts. By studying ancient primate species and their responses to ecological shifts, conservationists can draw parallels to contemporary species and their challenges. Fossil evidence and the study of extinct species, such as *Sivapithecus*, highlight how primates have historically adapted to changing climates, shifting landscapes, and evolving food sources (Almécija et al., 2021; Bhandari et al., 2018). Understanding these past adaptations allows conservationists to predict how modern primates might respond to similar environmental pressures today.

For example, archaeological findings from tool-use sites show how primates in the Nicobar Islands and Thailand have adapted their foraging strategies over time in response to fluctuating food availability (Luncz et al., 2017; Mazumder & Kaburu, 2020, 2021; Pal et al., 2018). In times of scarcity, tool use becomes more frequent, suggesting that the development of such behaviors is closely tied to ecological stressors (Mazumder, 2022). This indicates that primates' capacity for innovation is a key survival strategy, one that conservationists must factor into management plans. Ensuring that primates retain access to a diverse range of resources is essential, as their tool use may increase when other food sources become limited.

Furthermore, the archaeological record offers a long-term view of how ecological disturbances, such as habitat fragmentation or climatic shifts, have historically impacted species survival (Luncz et al., 2017). By tracing the extinction events and population declines of ancient primates, researchers can identify critical thresholds beyond which species are unable to recover. These findings underscore the importance of timely and proactive conservation interventions to prevent modern primates from crossing similar thresholds.

1.5.3 Integrating Behavioral Insights into Conservation Strategies

Conservationists increasingly recognize that behavioral adaptations, such as tool use, must be considered when developing strategies to protect endangered species. Traditional conservation approaches often focus on population numbers (Cosby et al., 2024), genetic diversity (Praveen Karanth, 2010), and habitat protection (Anitha et al., 2013), but they can overlook the importance of preserving species-specific behaviors that are crucial for survival (Behie et al., 2019). In the case of the Nicobar long-tailed macaques, tool use is not just a behavioral quirk; it is a key adaptive strategy that allows the macaques to access high-value food resources in their environment.

Conservation plans for these macaques must, therefore, include efforts to maintain the ecological conditions that allow for the continuation of tool use. This may involve protecting areas where suitable stones are available, ensuring that food resources such as nuts and mollusks are not overharvested by humans, and preventing habitat degradation that could disrupt foraging activities. By focusing on the preservation of behaviors as well as species, conservationists can help maintain the ecological integrity and cultural richness of primate populations.

Moreover, integrating the study of primate behavior into conservation strategies offers new opportunities for community engagement and education. Local communities in the Nicobar Islands, as well as tourists and broader conservation networks, can be made aware of the cultural significance of tool-using macaques. Promoting ecotourism centered around the observation of these unique behaviors could generate funding for conservation efforts while raising awareness of the need to protect the macaques' habitat.

1.5.4 Addressing Human–Wildlife Conflict and Conservation Challenges

One of the major challenges facing primate conservation in India is the growing conflict between human populations and wildlife (Jain, 2015). In regions where habitat loss and resource competition are severe, primates are increasingly coming into contact with humans, leading to crop raiding, property damage, and sometimes violent retaliations (Hill & Webber, 2010; Kaburu et al., 2019). For tool-using species like the Nicobar long-tailed macaques, human-wildlife conflict presents a unique challenge, as their behavioral traits may bring them into closer contact with human settlements in search of food (Mazumder & Kaburu, 2020; Mishra, et al., 2020; Pal et al., 2017).

Conservationists can address this issue by using insights from primate archaeology to design conflict-mitigation strategies. For example, understanding the types of resources that macaques seek when using tools can help in creating alternative food sources that draw them away from human crops. By preserving natural foraging areas and reducing resource competition, conservationists can minimize the negative impacts of human-wildlife interactions, ensuring the long-term survival of both primate populations and their cultural behaviors.

In conclusion, the study of tool use in primates, particularly the long-tailed macaques, offers profound insights into their cognitive evolution, adaptive strategies, and material culture. Through archaeological techniques such as excavation, artifact analysis, and experimental archaeology, researchers have been able to reconstruct the tool-using practices of long-tailed macaques, revealing the complexity of their behaviors and their cultural transmission across generations. The consistency in tool use, as evidenced in long-term tool-use sites, reflects the significance of these behaviors not only in the survival of the species but also in their cultural identity.

Looking ahead, there is great potential for advancing this field even further in India. Future research could incorporate cutting-edge techniques like 3D scanning and imaging to better analyze the micro-wear on tools, offering more precise insights into how they were used. In the future, should government regulations permit, the excavation of these sites, along with the collection and preservation of raw materials and artifacts, could provide valuable opportunities for advancing primate archaeological research in the region. Genetic studies may also provide a deeper understanding of how these behaviors are inherited, giving us a more holistic view of cultural transmission in primates. These innovations could unlock new dimensions in the study of primate archaeology, helping to clarify the links between behavior, environment, and culture.

Moreover, the conservation implications of this research are undeniable. By protecting the habitats of tool-using primates, we are not only safeguarding the species but also preserving the rich behavioral traditions that define their place in the ecosystem. As we continue to link past and present, this research will play an essential role in shaping future conservation strategies, ensuring that both primates and their remarkable cultural heritage endure. We need to keep up this inspiring work, as it contributes not only to science but to the preservation of these amazing creatures.

Note

* Detailed discussion on other tool-using primates can be found in Chapter 6 of this book.

2 Knowing the Islands and the Animal

The Andaman and Nicobar Islands, an archipelago steeped in natural beauty and cultural complexity, offer a fascinating perspective on ecological and social adaptation. The islands' history is rich with stories of indigenous communities who have long inhabited these lands, developing distinctive cultural practices and ecological knowledge that have enabled them to thrive in an isolated environment. Alongside these human histories, the islands are home to unique wildlife, including the Nicobar long-tailed macaque (*Macaca fascicularis umbrosus*), a primate that has adapted remarkably to the archipelago's challenging conditions. This chapter examines the interwoven histories of the islands, their people, and their wildlife, with a special focus on the developmental stages of the Nicobar long-tailed macaque. Tracing the macaque's life stages from infancy to old age, we explore how environmental pressures and social adaptations shape the macaques' behavior and survival strategies, providing broader insights into adaptation within insular and interdependent ecosystems. This contextual grounding is essential for interpreting the emergence, persistence, and archaeological visibility of macaque tool use and social learning, which are shaped by the islands' ecological constraints and human–animal relationships.

2.1 Andaman and Nicobar: The Islands and the Islanders

The Andaman and Nicobar Islands, an Indian Union Territory, stand as an ecologically rich archipelago located in the Bay of Bengal, extending in a long chain toward the equator. This archipelago comprises 836 islands (encompassing islands, islets, and outcrops), including the northern Andaman group and the southern Nicobar group, covering an area of approximately 8,249 square kilometers (Govt. of India, 2020). Its isolation and distinctive ecology make the Andaman and Nicobar Islands a rich study area for biodiversity, adaptation, and the symbiotic relationship between humans and their environment.

DOI: 10.4324/9781003614579-2

These islands are part of the Andaman–Nicobar accretionary ridge, a tectonically complex region arising from interactions between the Indian and Burmese plates (Bandopadhyay & Carter, 2017a). This dynamic geological environment has produced varied landscapes, including volcanic islands like Barren and Narcondam in the Andaman Sea and a striking physiographic division marked by the Ten Degree Channel. This 160-kilometer-wide channel separates the Andaman group from the Nicobar group, underscoring the archipelago's distinct geological identities and ecological differences (Govt. of India, 2020; ISFR, 2019). The region's isolation has fostered unique ecological niches and contributed to high rates of endemism among species such as the Andaman serpent eagle (*Spilornis klossi*), Nicobar megapode (*Megapodius nicobariensis*), Nicobar treeshrew (*Tupaia nicobarica*), and various flora like Andaman redwood (*Pterocarpus dalbergioides*) (Barstow, 2018; Thiollay, 1997, 1998).

The diverse blend of flora and fauna in the Andaman and Nicobar Islands is further complemented by its indigenous populations, including the Great Andamanese, Onge, Jarwa, Nicobarese, and Shompen. For centuries, these communities have sustained lifeways finely tuned to the island ecosystems, drawing on traditional knowledge to navigate the complex relationships between nature and external influences (Kar, 2019; Lobo, 2021; Rizvi, 1990). These relationships are grounded in the physical and ecological characteristics of the islands themselves, whose diverse landscapes and rich biodiversity shape both cultural practices and human–primate interactions. It is therefore essential to first outline the geography and ecological context of the archipelago before turning to its cultural heritage.

2.1.1 Geography and Ecological Characteristics

The Andaman and Nicobar Islands' geography includes mountainous terrain, dense forests, and vast coral reefs (Bandopadhyay & Carter, 2017b). The islands are divided into two major groups: the Andaman Islands and the Nicobar Islands. The Andaman Islands, situated further north, are generally larger and more elevated, with Saddle Peak reaching a height of 733 meters on North Andaman Island (Bandopadhyay & Carter, 2017a; Rodger & Panwar, 1988). The Nicobar Islands, in contrast, are smaller, with lower elevation and dense forest cover (Figures 2.1 and 2.2). This diverse terrain and ecosystem create distinct ecological niches that foster a high rate of endemism and specialization among various species (Bandopadhyay & Carter, 2017c; Director, 1995).

The islands experience a tropical monsoon climate, with average annual temperatures ranging from 23°C to 30°C and elevated humidity levels (Govt. of India, 2020). Monsoon rains, occurring from May to November,

Figure 2.1 Pristine Coastal Rainforest and Sandstone Escarpments of Great Nicobar Island.

recharge the islands' freshwater sources, although they can also lead to temporary flooding. The climate and its seasonal variations significantly influence resource availability for both humans and wildlife, including the endemic Nicobar long-tailed macaque, which demonstrates behaviors closely aligned with seasonal cycles.

The islands also face environmental challenges from climate change, particularly rising sea levels and the intensification of monsoonal cycles (Bellard et al., 2012). Coral reefs – essential for protecting coastal areas and supporting marine biodiversity – are threatened by warming seas, acidification, and human activities such as overfishing (Hoegh-Guldberg et al., 2007). Efforts to address these impacts include pollution control, sustainable fishing practices, and coral restoration projects, although these remain complex in implementation due to the islands' geographical isolation and limited infrastructure.

2.1.2 *Climate and Seasonality*

The Andaman and Nicobar Islands are notably vulnerable to climate change, confronting issues as mentioned above, such as rising sea levels and extreme weather events. This region experiences a tropical monsoon climate and high humidity year-round. The effects of climate change, such

Figure 2.2 Eroded Rocky Coastline and Uplifted Terraces of the Nicobar Archipelago.

as frequent extreme weather events, rising sea levels, and more, pose significant risks to the Andaman and Nicobar Islands (Bellard et al., 2012). Coastal mangrove forests, which play a crucial role in mitigating storm impacts and preventing erosion, are particularly sensitive to these changes (CIFOR-ICRAF, 2024; Hoegh-Guldberg et al., 2007). Additionally, coral reefs are increasingly subjected to frequent bleaching due to warmer ocean temperatures. These environmental fluctuations disrupt traditional livelihoods and resource availability, prompting the need for adaptive strategies that support both ecological health and cultural continuity.

The Nicobar long-tailed macaque, an endemic subspecies, has developed unique survival strategies, including foraging in coastal areas and timing its reproductive cycles to align with food availability, thereby maximizing offspring survival (Figure 2.3) (Mazumder & Kaburu, 2020; Narasimmarajan & Raghunathan, 2012; Pal et al., 2018). For indigenous communities, these environmental changes threaten traditional food sources and disrupt cultural practices, as shifting ecosystems alter resource availability.

Figure 2.3 A Young Adult Male Nicobar Long-Tailed Macaque (*Macaca fascicularis umbrosus*) Descends a Tree in Its Coastal Forest Habitat.

2.1.3 Biodiversity and Endemism

The Andaman and Nicobar Islands provide a compelling case study in biodiversity and the intersection of culture and environment. Their geographic isolation has resulted in a high degree of endemism, with various species uniquely adapted to the particular island ecosystems (Director, 1995). Notable endemic species include the Nicobar macaque, the Andaman serpent eagle, and the Nicobar megapode, along with distinctive flora such as the Andaman redwood and the Nicobar pigeon (*Caloenas nicobarica*). These islands boast a variety of habitats, including mangroves, tropical rainforests, and coral reefs, each supporting a diverse array of life forms (Figures 2.4 and 2.5) (Alappatt, 2017).

The unique ecological adaptations observed in species like the Nicobar long-tailed macaque, which demonstrates tool use – utilizing stones or hard surfaces to access hard-shelled foods such as coconuts – underscore the evolutionary responses to the limited and fluctuating resources available on the islands (Das et al., 2020; Mazumder & Kaburu, 2020; Pal et al., 2017). The coral reefs surrounding the islands serve as vital ecosystems, teeming with an array of marine life, including fish, mollusks, and sea mammals. These reefs are crucial not just for their biodiversity but also for their role in supporting local fisheries and the cultural practices of indigenous communities (Ameneshewa et al., 2023).

Figure 2.4 A Flowering Food Plant (*Melastoma malabathricum*) in the Nicobar Islands.

Figure 2.5 The Fruiting Stage of a Key Food Plant (*Melastoma Malabathricum*) in the Nicobar Islands.

Complementing this ecological richness are the islands' indigenous communities – the Great Andamanese, Onge, Jarwa, Nicobarese, and Shompen – whose distinct cultural traditions and subsistence practices have coevolved with, and remain deeply tied to, these environments (Cooley, 2014; Lobo, 2021; Macnaughtan, 2013; Rizvi, 1990; Sekhsaria & Pandya, 2010; Sharief & Panda, 2021; Shashi, 1994). These communities exemplify sustainable living practices that are finely attuned to the environmental rhythms of the islands. Notably, the Nicobarese and Shompen's seasonal resource use and the incorporation of ecological knowledge into their daily lives reflect an intricate understanding of resource management that promotes a balanced relationship between humanity and nature, contributing to both cultural and environmental resilience. Awaradi (2022) highlights that the fragile ecosystem is intricately linked to the cultural practices of indigenous groups, whose livelihoods depend on the land. Indigenous knowledge systems play a pivotal role in the conservation of these ecosystems; however, they are increasingly challenged (Arora, 2010; Awaradi, 2022; Bellard et al., 2012; Lobo, 2021). Coral bleaching, induced by rising sea temperatures, alongside issues such as overfishing and pollution, places significant pressure on these sensitive ecosystems. Conservation projects aiming to address these problems mostly focus on pollution control, sustainable fishing practices, and reef restoration efforts.

Here, Awaradi (2022) emphasizes the islands as 'ethnographic hotspots,' underscoring the need for urgent multidisciplinary studies to address the swift socio-environmental changes affecting indigenous lifestyles. Traditional systems are under siege from numerous external influences, such as tourism, development pressures, and the incursion of global markets. Moreover, the introduction of new policies – frequently designed without prior consultation with indigenous communities – can unintentionally disrupt sustainable practices (Ameneshewa et al., 2023; Dawson et al., 2024; Slocombe, 2000). Thus, collaborative research incorporating anthropology, ecology, and policy studies is essential to mitigate adverse effects while respecting indigenous rights and knowledge, safeguarding both the indigenous communities and their cultural heritage.

2.1.4 Impact of External Influences

The Andaman and Nicobar Islands, owing to their strategic location and rich natural resources, have long attracted external influences, shaping the islands' socio-cultural and environmental landscape. From early encounters with explorers to colonial rule and modern economic development, these influences have brought substantial transformations, often disrupting traditional ways of life and the natural environment (Krishnakumar, 2009; Sen, 2017). Today, as tourism and development continue to reshape the islands, the challenge lies in balancing economic growth with sustainable practices that respect and preserve the indigenous cultures and fragile ecosystems of the archipelago.

2.1.4.1 Colonial Era: Transforming Landscapes and Societies

The most profound and far-reaching external influence came with British colonization, which began in the 19th century (de Beaune, 2004). The British established a penal colony in the Andaman Islands in 1858, primarily using it to incarcerate Indian freedom fighters and criminals. The infamous Cellular Jail in Port Blair, known as *Kala Pani* (translating to 'Black Water'), became a symbol of colonial repression, where political prisoners faced isolation and harsh treatment (Sen, 2010). The penal settlements not only marked the region as a site of colonial dominance but also had lasting social impacts, as people from various regions and backgrounds were forcibly brought here, creating a demographic shift that resonates until today.

While the region had been known to the world since the era of Ptolemy, its unwritten history reflects deep connections between the indigenous Nicobarese and their environment, sustained primarily through activities like the copra trade (August, 1840; Govt. of India, 2022a). The colonial transformation began in earnest with the Danish settlement in 1756, marked by the establishment of a headquarters on Kamorta Island. However, this attempt at colonization was short-lived, as the Danes abandoned the islands by 1848, unable to maintain a foothold (Govt. of India, 2022a).

In 1869, the British formally annexed the Nicobar Islands, incorporating them into the administrative framework of British India. By 1870, a penal settlement was established on Kamorta Island, though it was discontinued by 1888 (Govt. of India, 2022b). The inclusion of the Nicobar Islands in the Chief Commissioner's jurisdiction of the Andaman and Nicobar Islands in 1871 signaled a formal administrative integration that would persist until independence (Govt. of India, 2022b). This integration came with significant changes: the imposition of colonial policies disrupted indigenous practices and introduced external influences that began altering the social and ecological landscapes.

The colonial period saw the introduction of plantation economies, primarily focusing on crops like coconut and rubber, which aligned with British economic interests (Malik, 2024; Sen, 2010). This shift, coupled with extensive logging for timber exports, led to deforestation and the degradation of ecosystems that had long supported the Nicobarese way of life (Sekhsaria & Pandya, 2010; Sharief & Panda, 2021). These environmental changes eroded the traditional self-sustaining systems that indigenous communities had maintained for generations (Chatterjee, 2020; Sen, 2010).

Furthermore, colonial policies marginalized indigenous groups, imposing alien systems of land ownership and resource management that disregarded their customary practices. Access to forests and coastal areas – crucial for traditional livelihoods – was curtailed, undermining the autonomy and ecological balance that had defined Nicobarese society (Chatterjee, 2020; Portman, 1896). The British administrative presence in the Nicobar Islands, epitomized by the appointment of the first Assistant Commissioner in the 1920s, entrenched new social hierarchies and introduced governance structures that often clashed with local traditions.

Thus, British colonialism not only transformed the physical landscape of the Nicobar Islands through deforestation and plantation agriculture but also disrupted the social fabric, displacing indigenous communities and redefining their relationship with the environment. These changes have left enduring legacies that continue to shape the islands' ecological and cultural narratives.

2.1.4.2 *Post-Independence Development and Changing Economies*

Following India's independence in 1947, the Andaman and Nicobar Islands were incorporated as a Union Territory. The Nicobar Islands, having briefly reverted to British control after World War II and Japanese occupation (1942–1945), became part of the newly independent nation on August 15, 1947 (Govt. of India, 2022a; Sen, 2018). During this period, John Richardson emerged as a significant figure in modern Nicobarese society, being consecrated as Bishop in 1950 and later representing the islands in Parliament as a nominated member, symbolizing a bridge between tradition and modernity (Govt. of India, 2022a; Sen, 2018).

Post-independence, the Indian government prioritized protecting the Nicobarese community, aiming to shield them from external exploitation. The enactment of the Protection of Aboriginal Tribes Regulation (1956) restricted entry to the Nicobar Islands, marking an early effort to preserve indigenous autonomy (Govt. of India, 1956). However, in the late 1960s, the government began settling non-tribal populations in the Nicobar Islands, mirroring initiatives in the Andaman Islands. Approximately 330 former defense personnel were resettled in de-reserved areas of Great Nicobar, and during the 1970s, Plantation Tamils were brought to Katchal Island for rubber plantation work. Population pressures also led to the relocation of 165 Car Nicobarese families to Little Andaman Island in 1973–1974 (Govt. of India, 2022b).

On 1 August 1974, the Nicobar Islands were designated as a separate administrative district, with Car Nicobar as the district headquarters, accommodating nearly half of the population. While Great Nicobar adopted modern governance structures with three Panchayats and one

Panchayat Samiti, the rest of the district preserved its traditional Tribal Councils (Govt. of India, 2022a), reflecting a blend of indigenous governance and administrative integration.

These developments occurred alongside broader trends in the Andaman and Nicobar Islands. New settlers, including refugees from partitioned areas and ex-servicemen, were introduced under government-sponsored rehabilitation schemes (Banerjee, 2024; Govt. of India, 2018; Menon & Krishnaswamy, 2024). This migration altered land-use patterns as agricultural expansion and infrastructure development often encroached upon forested and indigenous territories. While these settlers contributed to the economic and social diversity of the islands, competition over resources frequently marginalized indigenous groups.

Indian government policies have emphasized modernization and economic integration, leading to infrastructure development such as roads, ports, and airports (Lobo, 2021; Sekhsaria, 2024a). While these improvements facilitated connectivity and economic activity, they also fragmented ecosystems and introduced environmental pressures. Similarly, advancements in fishing, an economic mainstay for many coastal communities, have included mechanized practices that sometimes result in overfishing, threatening marine biodiversity.

In balancing development and conservation, the post-independence trajectory of the Nicobar Islands reflects a complex interplay between protecting indigenous identities, fostering economic integration, and managing the ecological impacts of modernization.

2.1.4.3 Tourism Boom – Opportunities and Environmental Strain

In recent decades, the natural beauty and biodiversity of the Andaman and Nicobar Islands have attracted increasing numbers of tourists (Sekhsaria, 2024b). Tourism has emerged as a primary source of income, particularly for urban centers like Port Blair and popular destinations such as Havelock Island. While tourism has undoubtedly contributed to economic growth, creating jobs and revenue streams for local communities, it has also generated environmental and social challenges.

Mass tourism leads to habitat encroachment, especially around coastal areas and coral reefs (Macdonald, 2024). Resorts, hotels, and other infrastructure developments to accommodate tourists have proliferated, often without sufficient environmental oversight (Sharma, 2020). Coral reefs, which attract divers and marine enthusiasts, have suffered from direct damage, as well as pollution and sedimentation caused by construction and human activity. The ecological footprint of tourism is evident in increased waste production, pollution, and habitat fragmentation, all of which threaten the sustainability of the islands' unique ecosystems (Biswas, 2024).

The rise in tourism has also influenced traditional lifestyles and consumption patterns. As local economies pivot toward catering to tourists, many residents find themselves shifting from subsistence livelihoods, such as fishing and small-scale agriculture, to service-based work, often in roles with limited job security and seasonal demand (Biswas, 2024; Kandhari Malhotra, 2024). This economic shift has mixed outcomes, providing income opportunities while also fostering dependency on fluctuating tourism markets (Kandhari Malhotra, 2024). Indigenous communities, in particular, face challenges as tourism imposes new cultural norms and expectations that sometimes clash with their traditional practices and values (Carr et al., 2016; Deborah Ramer, 2010).

2.1.4.4 *Contemporary Development – Infrastructure and Environmental Degradation*

Alongside tourism, the push for economic development has led to extensive infrastructure projects that further strain the islands' ecosystems (Biswas, 2024). Development initiatives aimed at enhancing connectivity and trade, such as new highways, airstrips, and shipping ports, have facilitated economic integration with mainland India but have also come at an environmental cost (Jain, 2015). Roads and infrastructure often fragment habitats, making it harder for wildlife to thrive, disrupting the migratory patterns of animals such as saltwater crocodiles and certain bird species (Jain, 2015). Moreover, construction activity contributes to soil erosion, which can have lasting impacts on the islands' fragile ecosystems (Jain, 2015).

Additionally, the extraction of natural resources to support construction projects has placed pressure on the islands' freshwater systems, which are already limited and vulnerable to salinity. The demand for building materials like sand and gravel has led to unregulated quarrying in some areas, exacerbating erosion and threatening coastal stability. The Andaman and Nicobar Islands' freshwater resources have also been impacted by tourism-related consumption, with local water sources sometimes diverted to meet the needs of hotels and resorts, limiting access for local residents and indigenous communities (EQUATIONS et al., 2008; Kar et al., 2016; Kannan et al., 2023).

2.1.4.5 *Social and Cultural Impacts – Shifting Identities and Erosion of Indigenous Knowledge*

The cultural implications of external influences on the Andaman and Nicobar Islands are complex. Due to tourism, indigenous communities, who have long sustained themselves through a deep understanding of their environment, now face the erosion of traditional knowledge and practices (Slocombe, 2000). The influx of mainland settlers and tourists has introduced new customs and consumer habits, which influence local identities, particularly among younger generations (Kumar, 1985; Zhuang et al., 2019). These cultural shifts are compounded by economic pressures that encourage indigenous people to adopt mainstream practices over traditional ones, leading to a gradual erosion of cultural heritage and language (Bromham et al., 2021; Ford et al., 2020; Kik et al., 2021).

Furthermore, the commodification of indigenous culture for tourism – such as organizing 'tribal tours' or displaying cultural artifacts – could sometimes lead to misunderstandings and oversimplified representations of complex social practices (Bandhu et al., 2024; Kar, 2019). Such portrayals can distort the public perception of indigenous communities, reducing their cultural practices to superficial experiences for tourist consumption and undermining their autonomy (Bandyopadhyay & Yuwanond, 2018; Chang, 2006; Westberg, 2023).

Efforts to preserve indigenous identity and heritage often clash with external demands for 'development' and 'modernization.' Some communities have resisted integration into tourism and the wage economy, seeking to maintain traditional livelihoods instead. However, without adequate support and recognition, these efforts are challenging to sustain, particularly as economic and environmental pressures mount.

2.1.4.6 Balancing Conservation, Development, and Indigenous Rights

The islands are home to numerous protected areas, including national parks and marine reserves, aimed at preserving their unique biodiversity (Arora, 2010, 2018; Forest Survey of India, 2019). However, the establishment of these protected zones often limits indigenous communities' access to traditional lands, impacting their way of life. Indigenous populations – including the Great Andamanese, Onge, Jarwa, Nicobarese, and Shompen – have long maintained a symbiotic relationship with the environment, developing subsistence practices aligned with ecological cycles and resource availability. Their cultural traditions promote sustainable interactions with the environment, reflecting a lifestyle that conserves resources and minimizes ecological disruption (Dawson et al., 2024; Priyadarshini & Abhilash, 2019; Slocombe, 2000).

A noteworthy example of this relationship is the Nicobar long-tailed macaque, which holds a significant role in Nicobarese and Shompen cultures (Macnaughtan, 2013). Among these communities, the macaque is regarded both as a kin figure and a competitor, with traditional beliefs discouraging harm and promoting coexistence (International Survival, 2024).

This respect fosters coexistence and acknowledges the macaque's contributions to seed dispersal and forest regeneration – key ecological functions that help sustain the islands' biodiversity. Such cultural perspectives contribute to a conservation ethic that aligns well with modern ecological principles.

While indigenous knowledge offers valuable insights for ecosystem management, current policies on protected areas often restrict indigenous access, sometimes causing friction between conservation goals and cultural practices (Ameneshewa et al., 2023; Slocombe, 2000). Integrating indigenous knowledge with scientific conservation efforts could enhance ecosystem management and support the sustainable use of resources (Priyadarshini & Abhilash, 2019). These policies have the potential to promote community-based conservation that aligns with indigenous values, thereby empowering local communities as stewards of their environment.

Furthermore, the social structure within the Andaman and Nicobar Islands reveals distinct gender dynamics, especially within indigenous communities. Drawing from insights in *A Gender Atlas of India*, progress and challenges in gender indicators on the islands – particularly in women's access to education, healthcare, and economic opportunities – reflect broader disparities (Kumar et al., 2018). Gender inequalities are often intensified in remote settings, where women face greater obstacles to accessing essential resources. Indigenous women, who are central to resource gathering and environmental stewardship, play a crucial yet frequently undervalued role in their communities.

Schuster et al. (2024) underscore the need for targeted interventions to address these challenges, especially given the gendered effects of climate change. As primary caregivers, women are disproportionately impacted by resource scarcity and natural disasters (Schuster et al., 2024). Promoting gender equity within conservation and development programs on the islands could enhance both social resilience and sustainable practices, supporting a balanced approach that values both environmental stewardship and community well-being.

2.1.5 Toward Sustainable Development – the Way Forward

The history of external influences on the Andaman and Nicobar Islands demonstrates the delicate balance between economic growth and environmental and cultural sustainability. As development continues, a shift toward sustainable and inclusive approaches is essential. Ecotourism, which promotes environmental conservation while providing economic benefits, has potential if managed with rigorous environmental standards and community involvement (Owen & Hanley, 2004). Policies that prioritize the preservation of ecosystems, support sustainable resource use, and recognize indigenous rights to land and cultural expression could foster a more harmonious balance (Walther et al., 2002).

The government and local authorities, in partnership with non-governmental organizations, could enhance conservation efforts by integrating indigenous knowledge into ecosystem management and encouraging community-led tourism models. Capacity-building programs for local communities could also empower them to take active roles in managing tourism and conservation, creating pathways for economic participation that align with cultural values.

Ultimately, the impact of external influences on the Andaman and Nicobar Islands underscores the need for policies that balance development with respect for environmental sustainability and cultural integrity. By fostering partnerships that prioritize ecological health and cultural resilience, the region can navigate the complexities of modernization in a way that honors its unique heritage and biodiversity. The challenge lies in ensuring that development does not come at the expense of the islands' irreplaceable natural and cultural assets but rather supports a sustainable future for all who call these islands home.

2.1.6 Human–Macaque Interactions – Coexistence and Ecological Balance in the Nicobar Islands

The Nicobar long-tailed macaque holds a unique place in the ecosystems and cultural landscapes of the Nicobar Islands. Endemic to the region, this macaque subspecies interacts closely with the indigenous Nicobarese and Shompen communities, which have historically maintained sustainable livelihoods intricately connected to their environment. These communities engage in subsistence activities that include fishing, foraging, and agriculture – all practices that depend on the islands' limited natural resources (Cooley, 2014; Macnaughtan, 2013; Sahay, 2021). The relationship between humans and macaques in this setting exemplifies a complex model of coexistence, where ecological interdependence and cultural reverence blend, creating a delicate balance that sustains both species (Gumert et al., 2013; Hill & Webber, 2010; Luncz et al., 2017; Wong & Candolin, 2015).

2.1.6.1 Ecological Roles and Shared Resources

The Nicobar long-tailed macaque is primarily frugivorous, feeding on the fruits of native trees, but its diet is notably adaptable, including foraging along coastlines for screw pines, coconuts, and other resources (Mazumder & Kaburu, 2021; Pal et al., 2017). This diet is aligned with the fluctuating availability of food across seasons, reflecting the macaque's resilience within the challenging island environment (Mazumder & Kaburu, 2021; Pal et al., 2017; Pal, et al., 2018). The macaque's foraging habits are ecologically significant, as they aid in seed dispersal, helping to regenerate forest vegetation and maintain biodiversity in the islands' fragmented landscapes.

However, the macaque's use of coastal areas for foraging often overlaps with the subsistence needs of local communities, leading to a unique competition for resources (Luncz et al., 2019; Sahay, 2021). During lean seasons or periods of environmental stress, such as after cyclones or heavy monsoon rains, both humans and macaques may rely more heavily on certain coastal areas for food. This can create tension, as the macaques forage in tidal zones and gather shellfish and crabs – resources that local people also depend on for food security. Yet, these interactions are mitigated by cultural practices and beliefs, which often frame coexistence with wildlife as a necessary part of life, fostering tolerance and even respect for the macaques despite occasional competition.

2.1.6.2 Cultural and Spiritual Significance of Macaques in Indigenous Communities

In indigenous Nicobarese and Shompen societies, the long-tailed macaque is more than just an ecological counterpart; it is a cultural symbol intertwined with local spiritual beliefs (Agrawal, 1969). Traditional views discourage harm to wildlife, reflecting an ethic of reverence for all life forms, which aligns closely with conservation principles. The Nicobarese people, for example, view macaques as both kin and competitors, embodying a dual role that acknowledges the animal's ecological importance as well as its occasional challenges to human activities (Agrawal, 1969). This view is reinforced by local taboos and spiritual beliefs that emphasize respect for non-human life, encouraging people to tolerate and protect wildlife, including the macaques.

Rituals and cultural practices often highlight this reverence. For instance, during community ceremonies and ancestral rituals, indigenous beliefs may portray the macaque as a creature that embodies resilience and adaptability, traits that are admired and respected within these societies. These beliefs contribute to a broader cultural ethic that not only discourages harming macaques but also underscores their role as essential contributors to the ecosystem. This cultural outlook fosters a symbiotic relationship, where both humans and macaques share the environment in a way that benefits the islands' overall ecological health.

2.1.6.3 Conservation Challenges and the Role of Indigenous Knowledge

The interactions between humans and macaques in the Nicobar Islands illustrate a model of coexistence that is both ancient and ecologically beneficial. However, external pressures – including habitat loss, climate change, and the expansion of tourism and development – pose challenges to this balance (Carr et al., 2016). Habitat fragmentation reduces the macaques' access to natural food sources, potentially increasing their dependence on shared resources with humans and intensifying competition (Ellwanger & Lambert, 2018; Hill & Webber, 2010). Additionally, as tourism grows, so does the risk of macaques becoming dependent on food provided by visitors, which can disrupt their natural foraging behaviors and lead to human–wildlife conflict (Mallapur, 2013; Maréchal et al., 2016).

Indigenous knowledge systems offer valuable insights into managing these challenges (Dawson et al., 2024; Singh & Sureja, 2008; Slocombe, 2000). By understanding the nuanced ways in which indigenous communities interact with and adapt to their environment, conservation strategies can be designed to support both human and macaque populations (Arora, 2010; Priyadarshini & Abhilash, 2019). For example, the Nicobarese people's use of traditional fishing practices, such as limiting fishing to specific seasons and avoiding overexploitation of certain species, reflects their understanding of marine resource conservation and the breeding cycles of aquatic life. Additionally, the Nicobarese people traditionally coexist with the Nicobar long-tailed macaque, adopting practices that minimize conflicts. They avoid harvesting fruit during peak macaque feeding times and leave portions of their orchards untouched, ensuring that both human and macaque populations can sustainably share resources. This reflects their deep ecological knowledge and commitment to maintaining harmony with local wildlife. These practices help maintain ecological balance and ensure resource availability for future generations. Incorporating these practices into modern conservation policies can improve ecosystem management and help prevent resource depletion and human-wildlife conflicts.

Furthermore, conservation programs that respect indigenous values and knowledge are likely to be more effective in the long term (Chatterjee, 2020). Policies that limit indigenous access to protected areas can unintentionally undermine the very ecosystems they aim to protect, as indigenous practices are often aligned with principles of sustainability and ecological balance. By involving indigenous communities in conservation planning and management, policymakers can create frameworks that support traditional knowledge while fostering biodiversity and habitat conservation. This approach aligns with a growing recognition in conservation science of the need to integrate indigenous perspectives to achieve sustainable environmental outcomes.

2.1.6.4 *The Path Forward – Balancing Conservation and Indigenous Rights*

The future of human–macaque interactions in the Nicobar Islands rests on the ability to balance conservation needs with respect for indigenous rights. Conservation efforts incorporating community-based strategies and prioritizing the coexistence of humans and wildlife can help maintain the delicate balance that has sustained the Nicobar Islands for generations (Behie et al., 2019; Cosby et al., 2024; Eichler Vercillo et al., 2022). Initiatives involving local people in habitat restoration, wildlife monitoring, and environmental education provide employment, foster community stewardship, and promote the conservation ethic already present in indigenous culture.

The human–macaque relationship in the Nicobar Islands provides a powerful example of how indigenous knowledge, cultural values, and ecological resilience can work together to support both biodiversity and sustainable livelihoods. The challenges of conserving this region while respecting indigenous rights require a multifaceted approach that combines scientific research, policy support, and community involvement. As development pressures increase, a sustainable path forward must balance conservation with cultural preservation, honoring the unique heritage of the Nicobar Islands and supporting a future in which both humans and wildlife can thrive together.

2.2 The Nicobar Long-Tailed Macaque – Adaptation and Social Structure in an Island Ecosystem

The Nicobar long-tailed macaque, a subspecies restricted to the Nicobar Islands (Great Nicobar, Little Nicobar, Katchall, and possibly in Kondul) (Figure 2.6), has evolved distinct behavioral, social, and ecological adaptations that enable it to thrive within the unique conditions of these islands (Ramakrishna et al., 2022). Isolated from mainland populations and constrained by limited resources and space, the Nicobar macaque's life stages – from birth through old age – are profoundly shaped by the demands of island ecology (Angst, 1975). This section explores each life stage, highlighting the macaque's adaptations in reproductive timing, social organization, and survival strategies. Additionally, this section also elucidates the intricate relationship between the macaque's behavioral ecology and the islands' environmental conditions. For clarity, the life-history stages discussed here broadly follow established macaque developmental categories: infancy (birth to ~1 year), juvenility (~1–3 years), adolescence (~3–5 years), adulthood (~5–15 years), and aging (>15 years), while recognizing that these boundaries vary with ecology, sex, and individual condition (Kalbitzer & Jack, 2018).

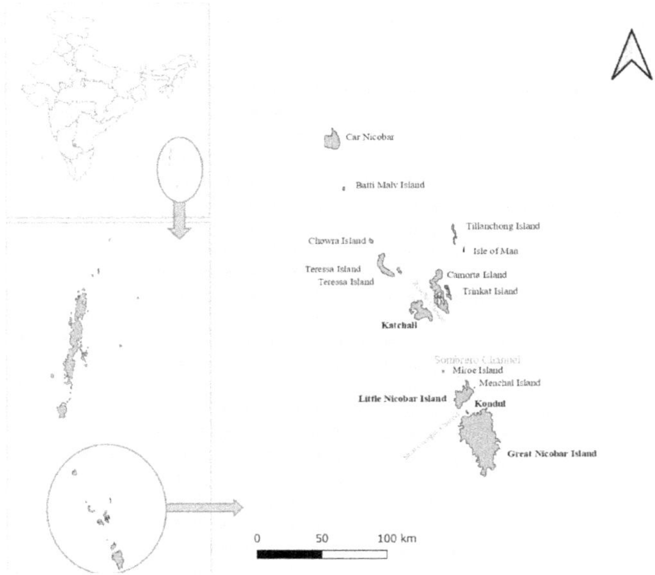

Figure 2.6 Map of the Andaman and Nicobar Islands, Highlighting the Distribution of the Nicobar Long-Tailed Macaque (*Macaca fascicularis umbrosus*) across Four Key Islands: Great Nicobar, Little Nicobar, Katchal, and Kondul.

2.2.1 *Birthing Seasons and Reproductive Timing – Adapting to the Rhythms of the Monsoon*

The reproductive timing of the Nicobar long-tailed macaque exemplifies a finely tuned adaptation to the seasonal rhythms of the Nicobar Islands, closely aligned with the monsoon season when food resources are at their peak (Pal, 2018; Pal et al., 2018). Observations reveal that approximately 71% of births within Nicobar macaque populations take place during the rainy months, indicating a strong correlation between reproductive events and resource availability (Pal et al., 2018). This synchronization likely enhances offspring survival in the resource-limited island ecosystem. Furthermore, this seasonal alignment provides a survival benefit by ensuring that sufficient resources are available to meet the increased nutritional needs of lactating females. However, recent studies emphasize that such narrow ecological coupling may also make the species vulnerable to climatic fluctuations: shifts in rainfall patterns and rising temperatures could disrupt the delicate balance between reproductive timing and resource abundance, potentially contributing to reduced fertility or lower birth rates over time (Gamalo et al., 2024; Velankar et al., 2024).

2.2.1.1 *Maternal Investment*

Although detailed behavioral observations on birthing in Nicobar long-tailed macaques are limited, studies on closely related *M. fascicularis* populations indicate that females exhibit considerable maternal investment during parturition, adapting their behaviors to maximize the survival chances of their offspring (Bercovitch, 2002). Females often select secluded or low-disturbance areas for giving birth, which minimizes predation risk and provides neonates with enhanced protection during their most vulnerable early weeks (Schülke et al., 2019; Trevathan, 2015). This maternal seclusion facilitates intensive grooming, serving both hygienic and bonding functions, thereby reinforcing the mother–infant connection that is critical for the neonate's development and early security within the social group (Bercovitch, 2002). Interbirth intervals average around 18 months, though these vary depending on reproductive success and maternal condition (Cawthon, 2006). Females often extend the period between births following successful rearing of an infant to independence, compared to shorter intervals after unsuccessful reproduction (de Jong et al., 1994; Van Noordwijk & Van Schaik, 1987, 1999). Such patterns underscore the substantial energetic and social investment required for infant survival, ensuring that adequate maternal resources are available before the next reproductive cycle.

2.2.1.2 *Ecological Synchronization*

The synchronization of birthing seasons among long-tailed macaques with periods of resource abundance represents an evolutionary adaptation to the fluctuating environment of the Nicobar Islands (Pal et al., 2018). Births peak between January and April, coinciding with the transition into the rainy season, when fruiting trees, young foliage, and other vegetation are most plentiful. The renewal of these food sources ensures that lactating mothers and weaning infants have access to abundant nutritional resources essential for early growth and survival (Melin et al., 2014; Prakash, 1958; Van Schaik & Brockman, 2005). By aligning reproduction with monsoon-driven resource cycles, females increase the probability of offspring survival, as juveniles benefit from sustained nutritional sufficiency during critical developmental phases (Van Schaik & Brockman, 2005). This ecological synchronization minimizes juvenile mortality and exemplifies a finely tuned reproductive strategy shaped by the selective pressures of island life, where resource availability and environmental stability fluctuate seasonally.

This reproductive pattern illustrates the significance of maternal care alongside the seasonal synchronization of birthing events. It highlights the evolution of species-specific life-history traits crucial for maintaining population stability in the face of ecological pressures characteristic of island ecosystems (Velankar et al., 2016). This relationship between reproductive strategies and environmental conditions provides insight into the adaptive mechanisms that enable species to thrive in isolated habitats.

2.2.2 *Infancy – Maternal Bonding and Early Development*

Infancy is a critical period of dependency and learning for the Nicobar long-tailed macaques. During this stage, infants rely heavily on maternal care, with physical proximity and grooming being primary modes of bonding and essential for social integration (Pal, 2018; Schülke et al., 2019). Although detailed behavioral studies on infancy in Nicobar long-tailed macaques are lacking, demographic and life-history data (Pal et al., 2018) suggest a birth environment where infants are likely highly dependent. Insights from studies on captive and other long-tailed macaques (Abney, 2023; Tsuchida et al., 2008) support the view that infancy in macaques is characterized by intensive maternal care and dependency, with mothers acting as primary caregivers and regulating infant access to others. Infants rely almost exclusively on their mothers for protection, nourishment, and social learning, making this phase essential for survival and social integration (Pal, 2018). The mother-infant bond is vital in an ecosystem where predation and resource scarcity are significant risks (Bercovitch, 1991; Fairbanks, 1996).

2.2.2.1 *Developmental Milestones*

In the absence of extensive behavioral data from Nicobar populations, studies on related *Macaca fascicularis* groups indicate that infancy involves strong maternal investment and learning through sustained physical and social contact. Macaques develop critical survival and social skills during infancy through close contact with their mothers (Maestripieri, 1994; Maestripieri et al., 2008). During this period, infants develop essential motor, vocal, and social skills that form the foundation for later group interactions (Bhan, 1992). Physical closeness – particularly through grooming – provides both comfort and opportunities for learning social cues (Carne et al., 2011; Dileep & Jose, 2014). Such early interactions foster attachment and trust, shaping the infant's understanding of group dynamics and social cohesion (Henazi & Barrett, 1999).

2.2.2.2 *Social Dependence and Learning*

The mother–infant bond serves as a primary gateway to socialization (Figure 2.7) (Maestripieri, 1994; Reyes & Klug, 2021; Trivers, 1972; Wang et al., 2021). Infants rely on their mothers for both physical support and

Figure 2.7 A Female Nicobar Long-Tailed Macaque Pulls Her Infant Close After a Brief Moment of Play, While a Juvenile Sits Nearby, Watchful and Calm.

social guidance (Arlet et al., 2020; Dunayer & Berman, 2018). Through observation and maternal supervision, they learn about dominance, affiliative behavior, and social hierarchies (Funkhouser et al., 2018; Maestripieri et al., 2008). Maternal oversight enables safe exploration of group interactions, preparing the infant for future independent participation (Fairbanks, 1996).

2.2.3 *Juveniles and Playgroups – Learning through Play*

Building on the general developmental trajectory of long-tailed macaques, the juvenile phase marks a critical period of increasing independence and social exploration (Anderson & Chamove, 1985; Luetjens et al., 2020; Trivers, 1972). This stage is characterized by the formation of playgroups, which provide an essential social context for learning through play. Playgroups often form around kin but also include unrelated individuals, offering diverse social partners for interaction (Ballesta et al., 2014; Sommer & Mendoza-Granados, 2010). Through these play interactions, juveniles acquire vital physical and social skills that prepare them for adult life, including the ability to navigate complex hierarchical relationships and form cooperative alliances.

a) **Role of Play in Development:** As in other *Macaca fascicularis* populations, play is a cornerstone of juvenile development, providing a safe context for experimentation, coordination, and learning (Fagen, 1982). Juvenile macaques engage in diverse play activities – such as mock aggression, chasing games, and foraging simulations – that contribute to both motor and cognitive growth (Pellis et al., 2019). These activities serve not only as physical practice but also as exercises in social negotiation and problem-solving.
b) **Motor Skill Development:** Activities like chasing and wrestling enhance physical agility, strength, and coordination (Tan, 2017). Foraging simulations help juveniles refine their grasping and manipulation skills, which are essential for acquiring food as they mature (Figure 2.8) (Albanese et al., 2022).
c) **Social Skill Development:** By participating in mock aggression and play fighting, juveniles practice key social behaviors in a low-stakes setting. They learn to interpret social cues, experiment with displays of dominance or submission, and adapt to changes in peer dynamics (Tan et al., 2018). Such activities allow young macaques to gain an early understanding of hierarchical structures, a central aspect of macaque society (Rapaport & Brown, 2008). Through these playful interactions, they learn boundaries, respect for authority, and the consequences of crossing those boundaries, all of which are integral to survival in their social world (Agostini & Visalberghi, 2005; Ballesta et al., 2014).

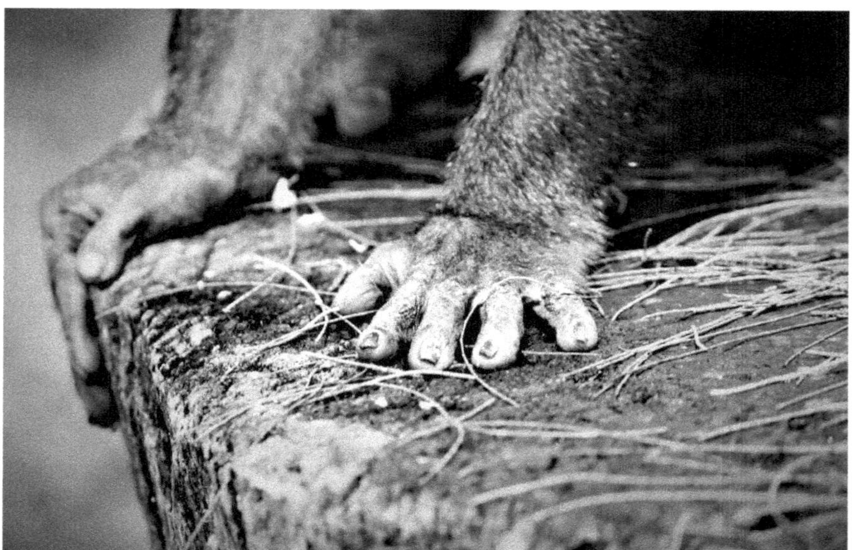

Figure 2.8 Close-Up of a Juvenile Macaque's Hands Placed on a Tree Stump While Standing.

2.2.3.1 *Formation of Social Bonds*

Playgroups act as microcosms of the adult social structure, where juveniles form bonds and alliances that often carry into adulthood. These early social bonds facilitate an understanding of group dynamics and establish friendships that may influence future alliances and support networks within the macaque society (Schülke et al., 2010).

a) **Development of Alliances:** Long-tailed macaques are well-known for their sophisticated foraging techniques, which include tool use, such as employing stones to crack open shellfish or coconuts, or manipulating objects to process food before consumption (Haslam et al., 2013; Mazumder, 2022; Mazumder & Kaburu, 2020). Juveniles are exposed to these behaviors during play, particularly in populations that engage in tool-assisted foraging (Tan et al., 2018). Through regular play, young macaques develop friendships that can mature into future alliances, which are essential in the highly social and often competitive nature of macaque society (Cawthon, 2006). Strong early bonds can profoundly influence social support networks, thereby affecting an individual macaque's access to resources, reproductive opportunities, and even their status within the troop (Micheletta et al., 2012; Schülke et al., 2010).
b) **Understanding of Group Dynamics:** The interactions that occur within playgroups closely reflect the dynamics of adult macaque society, effectively preparing juveniles to navigate intricate social relationships (Funkhouser et al., 2018; Krishna et al., 2006; Silk, 2007). By participating in these structured play environments, young macaques not only observe but also practice cooperative behaviors, conflict resolution, and reconciliation (Tan et al., 2018). They learn to identify reliable allies and potential competitors, establishing a solid foundation for lifelong social navigation.

Thus, play is a vital developmental mechanism that enhances juvenile macaques' physical agility, social understanding, and emotional resilience. Through these playful interactions, young macaques acquire the skills necessary to integrate into the adult hierarchy of their troop and cultivate the relationships that will support them throughout their lives. Consequently, the juvenile phase is not simply a time of unstructured play; rather, it is a critical period of learning that equips macaques with the essential tools for their future roles within the group.

2.2.4 *Adolescence – Social Navigation and Emerging Roles*

In the life cycle of the long-tailed macaque, adolescence marks a transformative stage of development. This phase is characterized by an increase in both social responsibility and interactions that directly affect an adolescent macaque's place in the group hierarchy (Cheng et al., 2023; Kuběnová et al., 2019). Adolescent macaques also shift their social roles and behaviors to prepare for adulthood as they undergo physiological changes (Bernstein, 2005). In this period, males and females begin to exhibit distinct social strategies, aligning with the different roles they will occupy within the group.

2.2.4.1 *Emergence of Dominance Behavior*

For adolescent long-tailed macaques, the emergence of dominance behaviors is a pivotal aspect of their transition to adulthood. Both male and female adolescents undergo important behavioral changes, yet they differ in focus and social approach:

a) **Male Adolescents and Competitive Displays:** Adolescent males start to assert themselves through competitive behaviors such as displays of physical strength, mock aggression, and minor conflicts with peers (Chandel, 1992). These behaviors are part of their journey toward establishing a rank within the male hierarchy, a position that will later influence their access to mates and resources. These displays may include chest-pounding, branch-shaking (by the use of resonance rods), and other dramatic behaviors meant to communicate strength and assertiveness (Cowlishaw & Dunbar, 1991; Georgiev et al., 2015; Mazumder & Kaburu, 2020). By engaging in these interactions, adolescent males practice essential dominance skills while also determining their social ranking. The outcomes of these early dominance encounters often forecast a male's standing in the adult hierarchy and directly impact his reproductive opportunities, given that higher-ranking males typically have priority access to females during mating seasons (Bercovitch, 1991; Schülke et al., 2010).
b) **Female Adolescents and Affiliative Behaviors:** Female adolescent monkeys, on the other hand, focus on affiliative behaviors that promote social harmony and cohesion (Borries et al., 1994; Chatterjee, 2003). Their social strategies involve grooming, playing with younger macaques, and forming alliances, especially with other females (Gumert, 2007). This behavior aligns with their eventual roles in nurturing and stabilizing group dynamics (Strier, 2021). Through these affiliative interactions, female adolescents form bonds that facilitate their social integration. These alliances with kin and non-kin alike will provide future support, particularly valuable when they are raising their own offspring (Sterck & Steenbeek, 1997; van Noordwijk & van Schaik, 1999). By honing their nurturing and affiliative skills

during adolescence, female macaques prepare for roles that contribute to group cohesion and enhance social stability, roles that are essential for the survival and success of macaque groups.

2.2.4.2 Complex Social Navigation

In addition to dominance and affiliative behaviors, adolescent long-tailed macaques engage in increasingly complex social navigation. This phase involves learning to balance competitive and cooperative interactions to optimize their social positions and prepare for adult responsibilities (Bernstein, 2005).

a) **Social Negotiation and Alliance Formation:** Adolescents frequently engage in alliances with peers and older group members. These alliances are fluid, shifting with social conditions, and provide adolescents with support during conflicts and influence within the group (Bernstein, 2005). Males, for example, may form temporary coalitions with peers to challenge older males, testing group boundaries while gaining experience in coalition-building, a skill they will use in future dominance conflicts (Chism & Rogers, 1997). Females, in contrast, use alliances to establish and maintain relationships, fostering the social bonds that will aid them in group cohesion and mutual support (Bercovitch, 1995). Female bonds are often more stable, as they are instrumental in the maintenance of social harmony within the troop.
b) **Adaptation to Group Dynamics:** Adolescent macaques learn to read social cues, navigate conflicts, and engage in conflict resolution, all of which are critical for managing the complexities of adult group dynamics. This adaptation is especially important in macaque societies, where alliances and social bonds can influence access to resources, protection, and reproductive opportunities (Overduin-de Vries et al., 2024). Long-tailed macaque adolescents are known for their sensitivity to social cues and group dynamics, learning how to handle conflicts and maintain their positions within the group (Creighton et al., 2021; Wong & Candolin, 2015). This stage of learning is particularly intense for males, who often disperse to join new groups upon reaching adulthood. By mastering social navigation and negotiation, adolescent males prepare themselves to establish favorable positions within new groups or to rise within their natal group's hierarchy. Female adolescents, who typically remain in their natal group, focus on reinforcing intra-group relationships. Their experience with social navigation at this stage is critical, as they are likely to interact with many of the same individuals throughout their lives.

Adolescence in long-tailed macaques is a developmental stage rich with learning and adaptation, as young macaques navigate their way through the intricate social structures of their group. Male adolescents practice dominance and competitive behaviors, essential for their future reproductive roles, while female adolescents engage in affiliative behaviors that promote group cohesion and prepare them for nurturing roles. Both males and females learn to manage social complexities, striking a balance between competition and cooperation that will serve them well in adult macaque society. Through these processes, adolescent macaques build the skills and alliances that lay the foundation for their future roles and social standing, highlighting adolescence as a critical phase for social navigation and role preparation.

2.2.5 Adulthood – Hierarchical Structure and Social Cohesion

Within the macaque tolerance–despotism continuum, Nicobar long-tailed macaques appear closer to the more tolerant end of *Macaca fascicularis* societies, particularly when compared to highly despotic species such as rhesus macaques (*Macaca mulatta*). In adulthood, long-tailed macaques fully integrate into the hierarchical social structure of their groups (Angst, 1975; Anuar & Ullah, 2011; Byrne & Bates, 2010; Pal, 2018). This phase is characterized by complex social interactions in which both dominance and cooperation play crucial roles in determining an individual's access to resources, mating opportunities, and social support. Adult males and females have distinct but complementary roles that uphold the group's social order and cohesion (Bercovitch, 1991; Bowles & Hammerstein, 2023). Observations suggest relatively flexible dominance interactions, frequent affiliative behavior, and reduced intensity of aggression, consistent with other coastal and insular *M. fascicularis* populations (Mitani et al., 2012). However, as with many macaque societies, tolerance is context-dependent and shaped by resource distribution, group composition, and human proximity.

2.2.5.1 Dominance and Reproductive Roles

Dominance in Nicobar long-tailed macaques is central to social organization, especially in resource allocation and reproductive success. High-ranking individuals, both male and female, experience significant advantages:

a) **Dominance in Resource Allocation:** High-ranking macaques generally have priority access to key resources, such as food and grooming partners. This preferential access is important in environments where food resources can be scarce, as it allows dominant individuals to maintain better nutrition and physical conditions, reinforcing their status (Canteloup et al., 2021; Sterck & Steenbeek, 1997). Dominant males secure priority in grooming and feeding locations, often positioning themselves in areas of high resource availability (Kuběnová et al., 2019; Sterck & Steenbeek, 1997). This control over resources enhances their physical well-being and signals their status to other group members, reinforcing their rank and social influence.

b) **Reproductive Roles and Success:**

- *Male Reproductive Strategies*: Dominant males enjoy higher mating success, as their elevated status affords them priority access to receptive females (Engelhardt et al., 2005, 2006; Mishra, et al., 2020). They often engage in displays of aggression or competition to deter subordinate males from mating, securing a reproductive advantage (Chism & Rogers, 1997). This priority access increases the likelihood of siring offspring, thus contributing to their reproductive fitness (Engelhardt et al., 2006).
- *Female Reproductive Advantages*: Dominant females often achieve higher reproductive success as well, though their advantage lies primarily in resource access (Borries et al., 1991; Cheng et al., 2023). High-ranking females benefit from their elevated social status by achieving reproductive maturity earlier, with a significantly reduced inter-birth interval (Cawthon, 2006) (Figure 2.9). For instance, high-ranking females have been observed to begin reproducing within 5.5 years, whereas lower-ranking females only start after 5.5 years, even when both reach sexual maturity at four years (Cawthon, 2006). This advantage is likely attributed to their prioritized access to resources, reduced stress from social subordination, and greater support within the social hierarchy, which collectively enhance their reproductive fitness. By having preferential access to food and social partners, dominant females can maintain better health, which directly benefits their offspring's survival rates (Anuar & Ullah, 2011). They can also better protect and nurture their young, leading to increased offspring survival and, in turn, a greater contribution to the genetic lineage of the group (Archie et al., 2014; Bercovitch, 1995).

2.2.5.2 *Social Cohesion*

Social cohesion is essential to the stability and survival of Nicobar long-tailed macaque groups (Byrne & Bates, 2010; Strier, 2021). Adult females play an especially important role in fostering these cohesive bonds, using

Figure 2.9 A Female Nicobar Long-Tailed Macaque (*Macaca fascicularis umbrosus*) Embraces Her Juvenile.

affiliative behaviors such as grooming to strengthen relationships across the group.

a) **Grooming Networks and Social Bonds:** Grooming is a cornerstone of macaque social life, and adult females are central to maintaining these networks (Henazi & Barrett, 1999; Schino & Aureli, 2008). Through reciprocal grooming, females reinforce social bonds, reduce tension, and stabilize group dynamics (Henazi & Barrett, 1999). These interactions create networks of cooperation and support that benefit the group as a whole (Wheatley, 1999). Grooming interactions are often exchanged based on rank, with higher-ranking individuals receiving grooming more frequently (Singh et al., 2006). However, subordinates also use grooming as a tool for alliance-building, providing grooming to dominant individuals in exchange for tolerance or protection. These relationships help reduce aggression and maintain group harmony, which is vital for the troop's survival in their challenging environment (Van Schaik et al., 2004).
b) **Hierarchical Order and Stability:** Affiliative behaviors among adults, including grooming, gentle touch, and proximity maintenance, reinforce the hierarchical order within the group (Thierry, 2007). These

behaviors are crucial for sustaining social stability, as they affirm the roles of individuals within the hierarchy and prevent the frequent conflicts that can arise in social animals (Hrdy, 1981; Schaub, 1996; Schino, 2007). By facilitating hierarchical stability, social cohesion supports group resilience in the face of external threats or resource scarcity (Cawthon, 2006). The cooperative alliances within the group, primarily facilitated by females, as they remain in their natal group, therefore, ensure that the group can function harmoniously, with limited disruption to access to resources or social support (de Jong et al., 1994; van Noordwijk & van Schaik, 1999).

c) **Cooperation and Group Survival:** Nicobar long-tailed macaques exhibit high social cooperation, which is vital in their natural environment. The cooperative aspects of their society, including mutual aid and forming alliances, provide security against external challenges, such as predators or competitors (Chapman & Rothman, 2009). Group cohesion, maintained by these social alliances, ensures that individuals within the group have access to support networks essential for physical and emotional resilience (Bercovitch, 1995; Chism & Rogers, 1997; Schülke et al., 2010). Male macaques, while primarily engaged in maintaining and asserting dominance, also contribute to group cohesion by participating in coalition-building, which can provide support during conflicts or in defense against threats (Aureli & Waal, 2000; Schülke et al., 2010; Thierry et al., 2008).

In adulthood, Nicobar long-tailed macaques are deeply integrated into their group's social structure, with distinct roles supporting individual success and group cohesion. Dominance hierarchies enable prioritized access to resources and mating opportunities, which is crucial for reproductive success, particularly for high-ranking males and females (Port et al., 2020; Sterck et al., 1997). At the same time, social cohesion, largely driven by female affiliative behaviors, sustains a cooperative framework that stabilizes group dynamics and supports group resilience. This combination of competitive and cooperative behaviors highlights the complexity of macaque social life, where both dominance and alliances are indispensable for individual and collective survival within the challenging environments of the Nicobar Islands.

2.2.6 Aging and Social Roles in Later Life

In long-tailed macaque societies, aging brings with it significant changes in social roles and behaviors. As primates grow older, they adapt to physical limitations by reducing their involvement in high-stakes interactions and dominance struggles (Strier, 2021; Sugiyama, 1994). However, aged macaques continue to play an important role within the group, contributing to social stability and cohesion. This stage highlights the adaptive nature of macaque social structures, where even physically declining individuals retain a degree of social influence and respect.

2.2.6.1 *Adaptations to Aging*

As long-tailed macaques age, they experience a gradual reduction in physical strength, agility, and overall stamina, which impacts their involvement in dominant and competitive activities (Newman et al., 2023; Rathke et al., 2022; Siracusa et al., 2022; Veenema et al., 2001). To adapt to these changes, elderly macaques modify their social interactions and rely more on established alliances to maintain their position within the group.

a) **Reliance on Established Alliances**: Older monkeys, particularly those with a history of high social rank, leverage long-standing alliances to maintain their place within the social hierarchy (Newman et al., 2023). These alliances, cultivated over a lifetime, provide a form of social security, allowing senior macaques to uphold their social standing without direct involvement in physical assertion or dominance struggles (Rathke et al., 2022; Silk, 2007). Elderly macaques engage in affiliative behaviors such as grooming and proximity maintenance with both peers and younger group members. These low-intensity interactions enable them to remain integrated and respected within the group, even as they withdraw from high-energy activities (Aureli & Waal, 2000).

For example, in Nicobar long-tailed macaques, an individual adult male named Picker, a handicapped older male, has been observed engaging in alloparent care with younger macaques (Figure 2.10) (Mazumder, 2018). This behavior, which includes protection, grooming, and social support, strengthened intergenerational ties and allowed the elder male to maintain relevance and influence within the group (Sinha, 1998; Whiten & Van De Waal, 2018). Such alloparent interactions reinforce social bonds and enhance group cohesion, as younger macaques benefit from the guidance and protection of experienced group members (Hector et al., 1989). Elderly females cultivate close connections with younger females and offspring, contributing to group stability and intergenerational continuity even as their physical involvement in competition wanes (Machanda & Rosati, 2020; Pavelka, 1990; Rathke & Fischer, 2021; Zhang et al., 2023).

Figure 2.10 Picker, a Resilient Male from the Study Group, Survived the Loss of Limbs to a Human Trap and an Eye in a Territorial Battle.

b) **Reduced Engagement in Dominance Conflicts:** As macaques age, many show reductions in physical strength and agility, which correspond with declines in dominance-related behaviors (Machanda & Rosati, 2020; Veenema et al., 1997). Instead of aggressive displays, older individuals often adopt more tolerant or affiliative roles, maintaining group cohesion through reduced competition and increased social tolerance (Bird & Bird, 1997; Rathke et al., 2022; Zhang et al., 2023). This reduction in dominant behavior is a practical adaptation, allowing aged macaques to conserve energy and minimize risk, while still maintaining a degree of respect and standing within the group (Van Noordwijk & Van Schaik, 1985). For former high-ranking males, the shift to less active social roles does not necessarily equate to a loss of status. Their historical dominance and established alliances often ensure that they remain respected figures, even as they reduce their physical involvement in group dynamics. This allows older males to 'retire' from intense competition while still playing a role in social organization.

2.2.6.2 *Group Responses to Mortality*

In primates, the death of a group member, especially an influential or dominant figure, has a noticeable impact on group dynamics and social structure (Gonçalves & Carvalho, 2019). Long-tailed macaques demonstrate behaviors indicating a degree of awareness and response to the absence of a deceased individual, reflecting the complex social structure of their society (Chin, 2024).

a) **Disruption and Reorganization:** When a high-ranking or socially connected macaque dies, the group often experiences a temporary period of disruption. Members may exhibit increased vigilance or changes in social behavior as they adjust to the new social reality. In the case of a deceased dominant male, there may be a period of increased competition among remaining males as they vie for higher ranks, resulting in a reorganization within the dominance hierarchy (Anderson, 2011; Thierry, 2007). Female groups, typically more stable and cohesive, may also respond to the loss of an elder female, particularly if she was a matriarchal figure. The absence of a key affiliative figure can lead to subtle shifts in social bonds and grooming patterns as individuals seek to re-establish connections and adapt to the group's new structure.
b) **Behavioral Responses to Loss:** Observations in non-human primates have shown that group members may display behaviors suggesting an awareness of the absence of a deceased individual (Zipple et al., 2021). In some cases, primates have been observed lingering near the body of a deceased group member or engaging in subdued behavior. These actions suggest a level of social recognition of the individual's role and the impact of their absence on group dynamics. The group's temporary shifts in cohesion following a death reflect the intricate social bonds that characterize primate society. These adjustments often involve increased grooming, affiliative behavior, and a restructuring of alliances, as group members work to re-establish stability. This capacity for social reorganization and the group's collective response to loss highlight the social complexity and interdependence within long-tailed macaque groups.
c) **Aging and Social Adaptation:** Aging in long-tailed macaques brings a shift from active competition to roles that emphasize social cohesion and stability. Elderly macaques adapt by relying on alliances and reducing engagement in dominance conflicts, which allows them to maintain social relevance despite the physical decline (Clancy et al., 2013; Machanda & Rosati, 2020; Rathke & Fischer, 2021). The group's response to the mortality of an elder, particularly a dominant or highly affiliative individual, reveals the depth of macaque social bonds. Temporary disruptions in group structure and increased affiliative behaviors following a death underscore the importance of individual roles and relationships in maintaining social stability. This later stage of life thus illustrates the adaptive nature of macaque social roles and the enduring value of senior members within their social communities.

2.2.7 Social Learning and Knowledge Transmission

Social learning is a cornerstone of macaque society, and in long-tailed macaque societies, social learning is a central mechanism for the acquisition of survival skills, social behaviors, and cultural norms (Aplin, 2016). This process allows younger macaques to learn from experienced group members, observing and practicing critical behaviors essential to their adaptation and survival. Older macaques, especially high-ranking individuals, play an active role in modeling behaviors, which facilitates the transmission of knowledge across generations and strengthens social bonds within the group (Coussi-Korbel & Fragaszy, 1995; Vale et al., 2017). Importantly, these processes do not imply intentional teaching or symbolic transmission, but rather rely on repeated exposure, tolerance, and social proximity, consistent with current understandings of primate social learning. This social learning and intergenerational knowledge transfer are essential to maintaining group cohesion and collective adaptability.

2.2.7.1 *Mechanisms of Knowledge Transmission*

In primate groups, knowledge transfer is primarily facilitated through observational learning, whereby younger individuals observe and mimic the actions of older, more experienced macaques (Mazumder, 2022; Mazumder & Kaburu, 2020, 2021; Tan et al., 2018). This transmission is particularly vital in resource-scarce environments, such as the isolated island habitats where some macaque populations live, and their survival hinges on specific, learned techniques for food acquisition.

a) **Tool Use and Foraging Skills**: One of the most notable examples of observational learning in long-tailed macaques is tool use, especially in populations inhabiting island environments with limited food resources (Mazumder, 2022). In these groups, adult macaques have been observed using stones to crack open hard-shelled foods, such as nuts (India) and shellfish (Thailand), a skill critical to accessing otherwise unavailable nutrients (Mazumder & Kaburu, 2020; Tan et al., 2015). Juveniles learn these techniques by closely observing adult macaques, often dedicating significant time to watching the process of selecting, positioning, and using stones effectively (Tan, 2017; Tan et al., 2018). This observational learning allows younger macaques to adopt proven methods more quickly and efficiently. Tool use in foraging not only enhances survival in environments with difficult-to-access resources but also illustrates the macaques' capacity for adaptability and innovation. Through this shared learning process, juveniles acquire skills essential for foraging and food acquisition, which are then incorporated into the broader set of group practices, ultimately enhancing group resilience in fluctuating environments.

b) **Innovation and Skill Refinement:** In addition to adopting existing foraging practices, juveniles sometimes make slight modifications or refinements to observed behaviors, contributing to innovation within the group. This learning environment encourages an adaptive approach to resource acquisition, allowing macaques to respond flexibly to changes in their habitat (Mazumder, 2022). The presence of skilled adult models provides a foundation for skill acquisition, while younger macaques, through experimentation and practice, occasionally introduce new techniques or variations, thereby enriching the group's collective knowledge base (Kummer & Goodall, 1985).

2.2.7.2 Mentorship and Cultural Transmission

Mature primates, particularly older females, function as key social models by providing repeated social exposure. Their routine participation in grooming, foraging, and social interactions allows younger group members to observe and gradually acquire these behaviors through social learning, without implying intentional teaching. (Falótico & Ottoni, 2014; Gumert et al., 2011; Kulik et al., 2015; Perry et al., 2017). This role is critical for the continuation of group-specific behaviors, which contribute to a shared social identity and group cohesion.

a) **Guidance in Foraging and Resource Management:** Older females often guide younger primates in locating and preparing specific types of food. This mentorship is not limited to passive observation; it includes direct engagement, such as demonstrating how to handle or process food items and leading juveniles to reliable foraging sites (van Schaik & Burkart, 2011). By actively transmitting knowledge of resource management, mature primates help ensure that younger individuals are equipped with the necessary skills to obtain sustenance within their ecological context (Rapaport & Brown, 2008). This hands-on guidance supports not only the individual's development but also the group's overall food security, as knowledge of diverse and effective foraging strategies is collectively maintained.
b) **Grooming and Social Etiquette:** Grooming, a central aspect of macaque social life, is another behavior passed down through mentorship. Grooming behaviors are not only practical but also serve to reinforce social bonds, reduce tension, and establish hierarchies within the group (Schino & Aureli, 2008). Older primates, particularly high-ranking females, model appropriate grooming practices, showing juveniles how to approach, groom, and respond to social cues within these

interactions (Schino, 2007). This transmission of grooming etiquette instills the group's social norms, teaching younger primates how to navigate complex social interactions, express affiliation, and respect social boundaries (Henazi & Barrett, 1999). Through these interactions, younger primates learn the subtleties of their social structure, including deference to higher-ranking individuals and the establishment of supportive alliances, both of which are essential to their future social integration.

c) **Cultural Transmission and Group Cohesion**: Social adaptation among non-human primates facilitates the transmission of cultural knowledge unique to each group (Angst, 1975; Aplin, 2016; Canteloup et al., 2021; Rapaport & Brown, 2008). This can include specific foraging techniques, grooming rituals, and even vocalizations that are distinctive to a particular population. These group-specific behaviors create a shared culture, promoting a sense of identity and continuity across generations (van Schaik & Burkart, 2011). Cultural transmission is particularly significant for macaques in isolated or challenging environments, such as those on the Nicobar Islands, where adaptability to local conditions is vital for survival (Mazumder, 2022). The continued practice and reinforcement of group-specific behaviors ensure that macaques maintain a cohesive and unified social structure, which is essential for collectively managing environmental challenges (van Schaik & Pradhan, 2003). The shared cultural practices established through social learning contribute to group cohesion by reinforcing a collective identity and fostering cooperative behaviors that benefit all members. These behaviors help macaques to navigate social dynamics, resolve conflicts, and maintain stability within the group.

Social adaptation and knowledge transmission play a fundamental role in the lives of long-tailed macaques, shaping their behavioral flexibility, social structures, and cultural identities. Through mechanisms such as observational learning, active mentorship, and the reinforcement of group-specific practices, younger macaques acquire the skills and knowledge essential for survival and social integration. This intergenerational transmission supports group cohesion and adaptability, enabling macaques to navigate their complex social environment and challenging ecological contexts collectively. Consequently, social adaptation serves as both a survival strategy and a means of cultural continuity, underpinning the long-term resilience of macaque societies.

In conclusion, the life stages of the Nicobar long-tailed macaque offer a remarkable view of adaptation within an insular ecosystem. From infancy through old age, these macaques demonstrate a finely tuned balance between ecological and complex social adaptation, highlighting the importance of a life-history approach to understanding their behavior. The unique conditions of the Nicobar Islands – including resource scarcity, isolation, and human interaction – have shaped their developmental strategies, social structures, and cultural practices. Macaques learn and transmit behaviors essential for navigating their environment and maintaining social cohesion through each stage of life. Social adaptation and intergenerational knowledge transfer emerge as key mechanisms for survival, allowing younger macaques to acquire vital foraging techniques, social skills, and group-specific behaviors with low risk of trial-and-error. Mature individuals, especially females, play essential mentoring roles that ensure the continuity of cultural practices, fostering resilience in the face of ecological and social challenges. This interplay of biological and social factors underscores the evolutionary significance of behavioral flexibility and points to the critical importance of conserving the macaques and their fragile habitat. In safeguarding their environment, we protect a species and the intricate cultural legacy that defines their society.

3 Food and Foraging Behavior of Nicobar Long-Tailed Macaques

The Nicobar long-tailed macaque (*Macaca fascicularis umbrosus*) exemplifies ecological and behavioral adaptations that have evolved in response to the challenges of life on the remote Nicobar Islands. Their foraging strategies, diet, social dynamics, and tool use reflect a nuanced interplay between environmental pressures and the species' innate behavioral flexibility. This chapter provides an in-depth view of their food and foraging behaviors, situating them within the broader context of primate ecology and insular adaptations.

3.1 The Islands' Ecology and Habitat Preferences

The Nicobar Islands, located in the southeastern Bay of Bengal, are an archipelago of immense ecological significance (Bandopadhyay & Carter, 2017a). Their geographic isolation and the region's tropical climate have contributed to the evolution of diverse habitats, ranging from tropical rainforests and mangroves to littoral zones and secondary-growth forests (Figures 3.1 and 3.2) (Alappatt, 2017; Forest Survey of India, 2019). The islands' ecological complexity is further accentuated by their unique climatic conditions, including high humidity, abundant rainfall exceeding 279 cm annually, and a pronounced monsoon season lasting from May to November (Kumar et al., 2012). These factors combine to create a dynamic mosaic of microhabitats that support a rich array of flora and fauna, including the Nicobar long-tailed macaque, an endemic subspecies that has adapted remarkably to this environment.

3.1.1 Ecological Features of the Nicobar Islands

The tropical rainforests of the Nicobar Islands are characterized by dense vegetation, with towering evergreen trees forming multiple canopy layers (Alappatt, 2017). This vertical stratification is critical for sustaining the islands' biodiversity, providing a range of niches for arboreal species

DOI: 10.4324/9781003614579-3

Figure 3.1 Coastal Rainforest Landscape of Great Nicobar Island.

Figure 3.2 Fringing Forest and Eroded Coastline of Little Nicobar Island.

such as the Nicobar long-tailed macaque. Key tree species in these forests include *Ficus* spp., *Mangifera indica* (mango), and *Artocarpus altilis* (breadfruit), which are integral to the macaques' diet. The understory is rich in shrubs, climbers, and epiphytes, contributing to the complex structure of these forests.

The mangrove ecosystems, primarily found along the islands' coasts and estuarine regions, play a crucial role in nutrient cycling and serve as important foraging grounds for the macaques, particularly during the dry season (Debnath, 2004). These habitats are dominated by mangrove species such as *Rhizophora* spp. and *Avicennia* spp., whose fruit and associated fauna, such as crabs and mollusks, provide supplementary food sources (Debnath, 2004; Roy et al., 2009; Thirumurugan et al., 2022).

Secondary-growth forests, which arise in areas disturbed by natural calamities like tsunamis, cyclones, or human activities, represent another critical habitat for the macaques (Narasimmarajan & Raghunathan, 2012; Porwal et al., 2012; Singh & Moeremans, 2017; Velankar et al., 2016). These forests are often dominated by fast-growing pioneer species that produce high volumes of fruits and seeds, making them an attractive foraging ground. The adaptability of the Nicobar long-tailed macaque to such disturbed habitats underscores its ecological resilience.

Littoral zones, encompassing sandy beaches, rocky shores, and intertidal areas, play a critical role in the foraging ecology of Nicobar long-tailed macaques, particularly during periods of resource scarcity (Figures 3.2 and 3.3). While these areas are not the macaques' primary habitats, they serve as crucial fallback zones when inland resources are limited due to seasonal fluctuations or ecological disturbances. The geological diversity of the Nicobar Islands significantly influences the availability and type of resources in these littoral zones.

The coastal zones of the Nicobar Islands are underlain by various rock types, including ophiolites, serpentinites, and sedimentary formations like limestone and sandstone (Desai, 2021). These are exposed along the shorelines due to tectonic uplift and erosion, creating a mosaic of microhabitats that support diverse intertidal flora and fauna, such as crabs, mollusks, and small fish, which also form a supplementary part of the macaques' diet (Bandopadhyay & Carter, 2017a, 2018; Ghosh et al., 2017). On rocky shores, the macaques have been observed using their dexterous hands to overturn rocks, exposing hidden prey such as crabs or snails (Gumert & Malaivijitnond, 2012; Mazumder, 2018). Their behavior is consistent with repeated use of predictable tidal conditions and the spatial distribution of prey, highlighting their ecological adaptability.

Sandy beaches provide access to stranded marine debris, such as dead fish or seaweed, which the macaques opportunistically consume. In some cases, macaques forage for intertidal organisms like barnacles and small

Figure 3.3 Intertidal Zone in Great Nicobar During High Tide.

crustaceans clinging to rocky substrates. These activities are facilitated by the accessibility of exposed corals and sandstone beds during low tides, which create flat surfaces ideal for foraging (Mazumder, 2018).

The use of littoral zones is not limited to food acquisition but also reflects the macaques' ability to exploit anthropogenic resources. Along the coastlines, they frequently scavenge discarded food items or debris brought in by ocean currents, demonstrating behavioral flexibility in adapting to human-influenced landscapes. However, such reliance on anthropogenic resources also increases the potential for human–wildlife conflict and poses conservation challenges.

The macaques' activities in littoral zones (Figure 3.4) have broader ecological implications. By foraging for intertidal species and scavenging along the shores, they contribute to nutrient cycling within these fragile ecosystems (Otani et al., 2020). Their movements and activities may also aid in seed dispersal along coastal areas, linking terrestrial and marine ecosystems.

Thus, the diverse geological features of the Nicobar Islands' littoral zones create a dynamic environment that supports the opportunistic foraging strategies of Nicobar long-tailed macaques. These behaviors highlight their ecological adaptability and underscore the importance of preserving coastal habitats, not only for the survival of this unique primate subspecies but also for maintaining the ecological balance of the region. Protecting

Figure 3.4 Same Coastal Stretch of Great Nicobar During Low Tide.

these zones from degradation caused by human activities, such as coastal development and pollution, is essential for ensuring the continued coexistence of macaques and the rich biodiversity of the Nicobar Islands.

3.1.2 Habitat Preferences of the Nicobar Long-Tailed Macaque

The habitat preferences of Nicobar long-tailed macaques are closely tied to resource availability. These primates are primarily arboreal, relying on the dense, mixed evergreen forests as their primary habitat. The extensive canopy cover of these forests provides safety from predators, ample foraging opportunities, and a network of trees for locomotion (Alappatt, 2017; Cawthon, 2006). Canopy continuity is especially vital for this species, as it allows them to traverse large areas while minimizing energy expenditure and exposure to ground-level threats.

Freshwater availability is another critical factor influencing habitat preference. The macaques exhibit a marked affinity for areas near streams, rivers, or natural waterholes, which serve as vital drinking water sources. These riparian zones also support lush vegetation, including fruit-bearing trees and insect-rich undergrowth, further enhancing their appeal.

Seasonal shifts in resource availability led to changes in habitat use (Melin et al., 2014; Pal, 2018). During the monsoon season, when the primary forests are abundant with fruits, flowers, and young leaves, the macaques predominantly remain in these dense areas. The availability of resources during this time supports larger group sizes and reduces the need for extensive movement (Pal, 2018; Pal et al., 2018). In contrast, the dry season compels the macaques to expand their range into secondary forests and even littoral zones to exploit alternative food sources (Berenstain, 1986; Mazumder, 2018). This seasonal movement demonstrates their adaptability and underscores the importance of maintaining connectivity between different habitat types to support their survival.

3.1.3 Role of Habitat in Foraging Strategies

The diverse habitats of the Nicobar Islands significantly influence the foraging strategies of the Nicobar long-tailed macaque (Lucas & Corlett, 1991; Mazumder, 2022). In the dense evergreen forests, the macaques primarily engage in frugivory, consuming a wide variety of fruits from trees such as *Ficus* spp., which produce figs year-round. Figs are particularly important because of their high carbohydrate content and consistent availability, serving as a keystone resource.

In secondary-growth forests (i.e., forests that have undergone regeneration following anthropogenic activities or natural disturbances, including agriculture, timber extraction, or catastrophic events), the macaques exhibit dietary flexibility by consuming fruits from pioneer species, seeds, and young shoots (Chokkalingam, 2001; Kassim et al., 2017; Mazumder, 2018). These forests often act as foraging hotspots when primary forest resources are depleted. Additionally, the open canopy of secondary forests provides opportunities for ground-level foraging, allowing the macaques to supplement their diet with insects and small vertebrates.

Mangroves and littoral zones (Figure 3.4) represent unique foraging habitats where the macaques display context-dependent foraging behaviors (Goutham-Bharathi et al., 2014; Mazumder, 2018). They have been observed feeding on crabs, mollusks, and other intertidal organisms in these areas (Mazumder, 2018; Welker, 1992). This behavior demonstrates repeated, practiced behaviors and adaptability, as foraging in these environments often entails managing challenging prey, such as slippery or hard-shelled organisms. The ability to utilize such diverse food resources across varying habitats underscores the macaques' behavioral plasticity and their capacity to exploit a broad spectrum of ecological niches. This adaptability aligns with findings by Luncz et al., 2017who documented how long-tailed macaques adjust their tool-assisted foraging strategies in response to resource availability, further highlighting behavioral flexibility in response to resource availability (Luncz et al., 2017).

The Nicobar Islands' diverse and dynamic ecosystems form the foundation of the Nicobar long-tailed macaque's survival. The species' habitat preferences, ranging from dense evergreen forests to coastal mangroves, reflect its remarkable adaptability and resilience. However, these habitats face growing threats, making their conservation critical for the macaques and the broader biodiversity of the islands. By understanding the Nicobar long-tailed macaque's ecological requirements and habitat use patterns, we can develop informed strategies to protect this unique primate and its environment for generations to come.

3.2 The Foraging Groups

Foraging is not merely a task of sustenance for Nicobar long-tailed macaques; it is frequently a group activity that reflects complex social dynamics, behavioral strategies, and ecological adaptations. In Nicobar populations, group sizes have been recorded between 25 and 56 individuals, with an average of around 36 (Boonkusol et al., 2018; Kumara et al., 2020). These macaques forage collectively, using extractive strategies to exploit hidden invertebrates and coordinate access to resources (Pal et al., 2017). Scramble competition and group-level foraging pressures also influence their ranging behavior (Mazumder, 2018; Velankar et al., 2024). A range of factors, including resource availability, predation risk, social structure, and environmental conditions, influence the size and composition of these foraging groups (Sterck & Steenbeek, 1997). This section delves into the characteristics of these foraging groups, their internal hierarchies, and the adaptive behaviors that allow them to thrive in the challenging environment of the Nicobar Islands.

3.2.1 Group Composition and Dynamics

While empirical documentation of this precise pattern in Nicobar long-tailed macaques is scant, primate foraging theory and macaque behavioral studies suggest that smaller family units may forage independently during periods of resource scarcity, while larger groups consolidate when food is clumped or abundant. For example, primate group living studies show that foraging efficiency and competition pressures drive subgroup formation under scarcity (Garber, 1987; Markham & Gesquiere, 2017). In macaques, changing patch utilization and movement strategies in response to food dispersion have been observed (Li et al., 2020). Moreover, in tool-using macaque populations, increasing group size and resource use lead to depletion pressures, necessitating adaptive foraging dynamics (Luncz et al., 2017).

In long-tailed macaques, the core of a foraging group typically comprises related females and their offspring, consistent with female-philopatric social organization. Adult males are generally transient; however, some may establish dominance within the group (Gumert, 2010). Among Nicobar long-tailed macaques, foraging behavior, including the use of cultivated resources like coconuts, varies across age and sex classes, with late juveniles, especially in urban troops, showing heightened use, and adult males typically consuming more than females outside this age group (Das et al., 2020; Velankar et al., 2023). Juveniles and subadults play an active role in group activities, often engaging in exploratory behaviors that contribute to context-dependent adjustments in learning and the discovery of new resources (Bernstein, 1976; Mazumder & Kaburu, 2020; Tan, 2017). The presence of an alpha male or female provides stability and leadership, particularly in coordinating group movements and defending against potential threats.

These groups display a clear dominance hierarchy, which dictates access to food and other resources (Mishra et al., 2020). Dominant individuals, often adult males or alpha females, are the first to access high-quality food sources, such as ripe fruits or protein-rich insects. Subordinate members must navigate this social hierarchy, often adopting rank-constrained responses to secure their share of resources.

3.2.2 Social Hierarchies and Resource Allocation

The hierarchical structure within foraging groups reduces overt competition and maintains group cohesion (Kazahari & Agetsuma, 2010). Dominant individuals assert their position through physical displays, vocalizations, and occasional confrontations, ensuring that access to resources follows an established order (Cawthon, 2006). This structure benefits the group by minimizing chaotic conflicts and promoting efficient resource use.

However, dominant members' monopolization of prime food sources often forces subordinate individuals to adapt their foraging strategies (Grueter et al., 2016; Mazumder & Kaburu, 2021). These adaptations in other primates are particularly evident in females with infants and juveniles, who may be unable to compete with higher-ranking group members directly (Sterck et al., 1997; Trivers, 1972).

3.2.3 Strategic Foraging by Females and Juveniles

Females with dependent offspring often employ foraging adjustments to compensate for their lower rank (Ram et al., 2003). One common behavior in long-tailed macaques is 'opportunistic following,' where the females closely trail dominant individuals to access discarded food or leftovers (Mazumder, 2018). This approach allows them to benefit indirectly from the dominant individuals' access to high-quality resources.

Juveniles exhibit unique foraging behaviors driven by curiosity and the need to learn. They frequently engage in exploratory foraging, sampling a diverse array of foods, including unfamiliar fruits, leaves, twigs, and even small invertebrates (Mazumder & Kaburu, 2020). This exploratory phase is critical for developing the dietary breadth and foraging skills necessary for survival in the complex environment of the Nicobar Islands. Studies on long-tailed macaques outside the Nicobar Islands indicate that juveniles acquire foraging skills primarily through observation and repeated exposure to older group members, with techniques becoming more refined over time (Tan et al., 2018). Importantly, these processes do not imply intentional teaching or imitation, but rely on repeated exposure, tolerance, and social proximity, consistent with current debates on primate social learning.

3.2.4 Foraging, Food Sharing, and Social Bonds

While competition is inherent in group foraging, Nicobar long-tailed macaques display context-dependent cooperative behaviors that not only enhance resource acquisition but also reinforce social bonds (Mazumder & Kaburu, 2021). These cooperative interactions are particularly notable in their food-sharing behavior, an infrequently documented pattern among macaques, highlighting elevated social tolerance in this population (Mazumder & Kaburu, 2021).

Group foraging often involves collective efforts to locate and access food sources. For instance, individuals coordinate their movements to fruiting trees, with dominant individuals often leading the group to known resource-rich areas (Mazumder & Kaburu, 2021). Once a food source is located, foraging is not a purely individualistic activity. The macaques demonstrate tolerance and cooperation, with group members taking turns accessing high-value resources or sharing feeding sites to ensure broader group benefits (Mazumder & Kaburu, 2021).

3.2.4.1 Food Sharing: an Unusual Behavior in Macaques

Food sharing, less frequently documented among macaques than among apes, is a particularly intriguing aspect of Nicobar long-tailed macaques' behavior. Observations have documented instances where dominant individuals, including alpha males or high-ranking females, share food with subordinates, juveniles, or even unrelated individuals. This behavior often involves sharing high-quality or scarce resources, such as ripe fruits, processed food such as coconut meat, or insects (Mazumder & Kaburu, 2021).

The motivations behind food sharing in these macaques appear to be multifaceted. It is associated with the reinforcement of social bonds, reducing conflict, or ensuring group cohesion (Mazumder, 2018). For example, mothers frequently share food with their offspring to provide immediate sustenance, an optimal energy source, and familiarize the young with diverse food sources. Juveniles, in turn, reciprocate by sharing their discoveries with peers or older group members, fostering social learning and collective foraging success.

Food sharing also occurs during periods of resource scarcity, when equitable distribution may enhance group survival. In these instances, dominant individuals have been observed tolerating close foraging by subordinates, allowing them to access leftover food or directly offering portions of their finds or processed food (Mazumder & Kaburu, 2021). This behavior reduces competition and ensures that even the least dominant members of the group can sustain themselves, thereby strengthening the group's overall stability and resilience.

3.2.4.2 Grooming Following Foraging Activities

Grooming is discussed here in its affiliative and tension-reduction functions rather than as a ritualized or normative behavior. Grooming sessions frequently follow group foraging activities, serving as a crucial mechanism for maintaining social cohesion. Grooming reduces tension arising from competition over resources, reinforces alliances, and fosters a sense of community within the group (Gumert, 2007; Mishra, et al. 2020). Individuals who engage in food sharing often receive grooming, suggesting that these interactions are reciprocated through affiliative behaviors (Mazumder, 2018).

In larger troops, where the potential for conflict is greater, grooming is more vital in mitigating disputes and building trust (Carne et al., 2011; Henazi & Barrett, 1999). In long-tailed macaques, grooming appears to be directed more toward high-ranking females, suggesting that subordinates may use grooming to gain favor or social benefits. However, dominant individuals themselves do not appear to use grooming to maintain status or bonds, as grooming was not more frequently initiated by them, nor were they central in the grooming network. Instead, middle-ranking individuals played a more active role in grooming distribution, indicating a more nuanced and possibly less steep dominance-based social strategy in this isolated population (Mishra et al., 2020). Subordinate individuals were observed directing grooming toward high-ranking females, which may reflect efforts to maintain favorable social relations; however, no direct link was found between grooming and proximity, suggesting that grooming may not necessarily translate into increased foraging opportunities or social tolerance in this species (Cawthon, 2006; Gumert, 2007, 2010).

3.2.4.3 Food Sharing and Social Hierarchies

The interplay between food sharing and social hierarchies in Nicobar long-tailed macaques highlights a structured social organization shaped by dominance and tolerance (Mazumder & Kaburu, 2021). High-ranking individuals often control access to premium food sources but exhibit a surprising degree of tolerance toward sharing (Mazumder & Kaburu, 2021). This behavior likely reflects the balance between maintaining dominance and fostering group harmony.

Subordinates typically approach dominant individuals using non-aggressive, rank-appropriate behaviors such as proximity, vocalizations, or submissive postures (Mazumder & Kaburu, 2021). The cooperative foraging and food-sharing behaviors of Nicobar long-tailed macaques illustrate the deep interconnection between their ecological strategies and social dynamics (Hohmann, 2009). By tolerating close foraging, sharing food, and engaging in post-foraging grooming interactions, these primates enhance their resource acquisition, reinforce social cohesion, and reduce intragroup tension. These behaviors underscore a structured social organization shaped by dominance, tolerance, and repeated encounters with the ecological challenges of their insular habitat. Understanding these dynamics provides valuable insights into the species' survival strategies and the importance of preserving their complex social and ecological systems.

3.2.5 Predation Risk and Group Size

Foraging group size may be influenced by perceived predation risk. Direct quantitative data linking group size and predation risk in Nicobar macaques are currently limited; therefore, this section draws on comparative primate literature. Observations discussed here reflect perceived risk inferred from habitat openness, proximity to humans and domestic dogs, and vigilance behavior, rather than measured predation rates or experimental comparisons. Larger groups offer safety in numbers, allowing individuals to spend more time foraging and less time being vigilant. In the dense forests of Southeast Asia, potential predators include raptors, snakes, and, in some areas, dogs (feral as well as guard dogs) (Mazumder, 2018; Riley et al., 2015).

Vigilance behaviors within foraging groups are often distributed, with dominant individuals frequently assuming the role of sentinels (Bednekoff, 2015). These sentinels position themselves in elevated locations or on the periphery of the group, scanning for threats and issuing alarm calls when necessary. This division of labor enhances the group's ability to detect and respond to potential dangers, allowing subordinate members, including juveniles and lactating females, to focus on foraging.

3.2.6 *Flexibility in Group Dynamics*

The flexibility of Nicobar long-tailed macaque foraging groups is a testament to their adaptability. The ability to adjust group size and composition based on environmental and social factors is crucial for survival in the resource-scarce and unpredictable conditions of the Nicobar Islands. For example, during periods of resource scarcity, groups may split into smaller units to reduce competition and increase foraging efficiency. Conversely, when resources are abundant, larger groups may form to maximize the exploitation of food sources while enhancing protection against predators (Garber, 1987; Markham & Gesquiere, 2017; Mazumder, 2018; Pyritz et al., 2013).

This fission-fusion dynamic also allows individuals to form temporary associations with other groups or solitary males, particularly in high-resource areas. Such interactions provide opportunities for genetic exchange, the sharing of information about food sources, and the reinforcement of broader social networks.

3.2.7 *Ecological Implications*

The social and foraging behaviors of long-tailed macaques have significant ecological implications. By consuming a wide variety of fruits, macaques play a critical role as seed dispersers, contributing to the regeneration of the islands' forests (Lucas & Corlett, 1998). Their preference for ripe fruits ensures that seeds are dispersed at the optimal stage for germination (Hohmann, 2009). Moreover, their movement across different habitats facilitates the spread of plant species, enhancing genetic diversity and promoting ecosystem resilience.

Insects and small invertebrates consumed by the macaques also serve as a link in nutrient cycling, highlighting the macaques' role in maintaining ecological balance (Hadi et al., 2007; Hasan et al., 2023; Wheatley, 1989). The selective foraging strategies employed by different group members further underscore their adaptability and the complex interplay between their social dynamics and ecological function.

The foraging groups of Nicobar long-tailed macaques exemplify the intricate relationship between social structure, ecological adaptability, and survival. From their hierarchical dynamics and cooperative behaviors to their responses to environmental challenges, these macaques highlight the importance of group living in navigating the complexities of their insular habitat. Understanding the nuances of their foraging strategies provides valuable insights into primate behavior and underscores the need for targeted conservation efforts to protect this subspecies and its fragile ecosystem.

3.3 Folivores or Insectivores – The Dietary Adaptations of the Nicobar Long-Tailed Macaque

Dietary patterns described here integrate field observations, published studies, and qualitative ecological surveys; quantitative proportions for all dietary categories are not yet available. The Nicobar long-tailed macaque exhibits remarkable dietary flexibility, which is a key factor in its survival within the diverse and often challenging ecosystems of the Nicobar Islands. As an omnivorous species, this macaque's diet encompasses a broad range of food sources, including fruits, leaves, flowers, seeds, and insects (Figure 3.5) (Mazumder, 2018; Mazumder & Kaburu, 2021; Pal et al., 2017). This dietary versatility enables the macaques to adapt to seasonal and ecological fluctuations, ensuring their survival in an environment characterized by resource variability (Mazumder & Kaburu, 2020). This section explores the dietary patterns of the Nicobar long-tailed macaques, focusing on their frugivorous tendencies, folivorous adaptations, and insectivorous behaviors. It also examines the ecological and physiological implications of these feeding strategies.

3.3.1 Frugivory – Fruits as a Staple

Fruits constitute one of the primary components of the Nicobar long-tailed macaque's diet, particularly during periods of high fruit availability. The

Figure 3.5 A Nicobar Long-Tailed Macaque (*Macaca fascicularis umbrosus*) Forages on Ripe Forest Fruits of *Melastoma Malabathricum*.

dense evergreen forests of the Nicobar Islands provide an abundance of fruit trees, with species such as coconut (*Cocos spp.*), figs (*Ficus spp.*), mangoes (*Mangifera indica*), and breadfruit (*Artocarpus altilis*) forming the cornerstone of their dietary intake (Mazumder, 2018). These fruits are rich in sugars, vitamins, and other nutrients, making them an energy-dense food source for the macaques (Hussain et al., 2024; Maldonado-Celis et al., 2019).

The macaques exhibit a preference for ripe fruits, which are not only sweeter and more palatable but also easier to digest due to their reduced fiber content (Kassim et al., 2017; Lucas & Corlett, 1991; Reinegger et al., 2023). Their foraging behavior during fruiting seasons often revolves around locating and consuming these high-energy resources (Das et al., 2020). Troops coordinate movements to areas with fruiting trees, sometimes covering substantial distances to exploit these ephemeral food sources.

The ecological role of the macaques as seed dispersers is significant (Lucas & Corlett, 1998). By consuming fruits and excreting the seeds at various locations, they facilitate the regeneration of forest vegetation and contribute to the ecological health of their habitat. Their preference for particular fruit species also influences the spatial distribution of these plants, underscoring the symbiotic relationship between the macaques and their environment.

3.3.2 Folivory – Adapting to Scarcity

During lean fruiting seasons, when fruits are scarce or unavailable, long-tailed macaques shift their dietary focus to leaves, flowers, and seeds (Cawthon, 2006). This transition to a folivorous diet highlights their dietary plasticity and their ability to exploit alternative food sources in response to resource scarcity. Periods of food scarcity refer to seasonal phases of reduced fruit availability, particularly during late dry seasons or following environmental disturbance, inferred from phenological observations, rather than quantified nutritional shortfall.

Although less energy-dense than fruits, leaves provide a reliable and abundant food source (Lambert, 2013). The macaques display a preference for young, tender leaves, which are higher in nutrients and easier to digest than mature, fibrous leaves (Coiner-Collier et al., 2016; Matsuda et al., 2017). Commonly consumed species include leaves from mangrove trees and other native flora of the Nicobar Islands.

Consuming leaves poses certain physiological challenges, as they are often high in fiber and secondary compounds such as tannins and alkaloids (Sharma et al., 2021). However, the macaques possess adaptations that enable them to process these less palatable food items effectively (Li et al., 2022; Zhang et al., 2022). Additionally, macaques exhibit behavioral and physiological adaptations to consume suboptimal food sources during periods of scarcity. For instance, during the lean dry season, they switch to consuming young leaves and seeds that are high in tannins and secondary metabolites. Their saliva contains tannin-binding proteins that mitigate the bitterness and toxicity of such compounds, while their molar morphology facilitates efficient mastication of fibrous plant matter (Prinz & Lucas, 2000). Additionally, prolonged chewing increases the surface area for enzymatic digestion, optimizing nutrient extraction from tough plant materials (Bennick, 2002). This combination of behavioral flexibility and physiological specialization allows macaques to maintain dietary breadth and survive in environments with fluctuating resource availability (Rothman et al., 2009).

Flowers and seeds also play a supplementary role in the macaques' diet during periods of fruit scarcity. The macaques are known to consume the nectar-rich flowers of certain trees, which provide an important source of energy, particularly during the dry season when other resources are limited (Garber, 1987). Seeds, although less commonly consumed, are accessed through the macaques' adept foraging techniques, such as cracking open seed pods or extracting seeds from fruits (Velankar et al., 2023).

3.3.3 Insectivory

A Protein-Rich Alternative: Insects form a critical component of the long-tailed macaque's diet, particularly during periods of resource scarcity or for specific physiological needs, such as lactation or juvenile growth (Ramakrishna et al., 2022; Sha & Hanya, 2013a; Uehara et al., 2021). The consumption of insects provides a rich source of protein, fats, and other essential nutrients that complement the carbohydrate-heavy diet derived from fruits and leaves. The macaques employ a variety of foraging techniques to capture insects, showcasing their behavioral flexibility in prey acquisition (Mazumder, 2018; Mazumder & Kaburu, 2020, 2021; Pal et al., 2017). These techniques include:

- **Flipping Over Leaf Litter:** Searching for ground-dwelling insects such as beetles, ants, and termites.
- **Peeling Bark:** Exposing hidden insects, particularly termites, caterpillars, and larvae, that reside beneath tree bark.
- **Probing Crevices:** Using their dexterous fingers to extract insects from narrow spaces, such as cracks in tree trunks or between rocks.

Commonly consumed insect species include termites, beetles, caterpillars, ants, and grasshoppers (Mazumder, 2018). The macaques clearly prefer nutrient-dense prey, often expending significant effort to capture protein-rich insects (Mazumder, 2022; Velankar et al., 2023).

Insectivory is particularly vital for lactating females and juveniles, who have higher nutritional demands (Altmann & Samuels, 1992; Perwitasari-Farajallah et al., 2022). The protein and fat derived from insects support milk production for lactating females, while for juveniles, these nutrients are crucial for growth and development (Hinde et al., 2009). The incorporation of insects into the diet also underscores the macaques' role as omnivores capable of utilizing both plant and animal resources.

3.3.4 Dietary Plasticity and Behavioral Adaptations

The ability to toggle between frugivory, folivory, and insectivory is a hallmark of the Nicobar long-tailed macaque's adaptability. This dietary plasticity reflects not only their ecological flexibility but also their capacity to adjust behavior through experience in response to changing environmental conditions. For instance, during extreme resource scarcity, the macaques have been observed exploiting unconventional food sources, such as crustaceans, mollusks, intertidal organisms, and even coconuts that were washed to the shores in coastal zones (Botting, 2021; Cawthon, 2006; Mazumder & Kaburu, 2020). These behaviors highlight their opportunistic foraging strategies and ability to adapt to diverse habitats.

Behavioral observations also indicate a degree of context-specific social and ecological adaptations to dietary choices. Juveniles often become familiar with available food sources through observation and repeated exposure to skilled group members, gradually acquiring the skills necessary to exploit a wide range of resources (Tan et al., 2018). Repeated social exposure may facilitate familiarity with food resources, enabling the group to adapt to fluctuating resource availability.

3.3.5 Ecological Implications of Dietary Choices

Nicobar long-tailed macaques exhibit remarkable adaptability in their foraging strategies, relying on a diverse range of plant species to meet their dietary requirements. Ecological surveys have identified numerous plant species that form the core of their diet, reflecting their ability to exploit resources across varied ecological zones (Mazumder & Kaburu, 2021).

Prominent among their food sources are fruit-bearing trees that provide high-energy sustenance essential for survival. Species such as *Neolamarckia cadamba* (kadamba tree or burflower tree), *Artocarpus sp.* (breadfruit and jackfruit), and *Tamarindus indica* (tamarind) offer nutrient-rich fruits and seeds. Coconut palms (*Cocos* sp.) and mango trees (*Mangifera andamanica*) are also vital, producing energy-dense fruits that are integral to the macaques' diet. Banana plants (*Musa* sp.) contribute significantly with versatile food options, including fruits, flowers, and stalks.

During periods of fruit scarcity, the macaques exhibit dietary plasticity by incorporating alternative resources. *Pandanus* sp. (screw pine or pandanus) provides both fruits and leaves, while *Melastoma malabathricum* (Malaysian melastoma or purple rose) and *Ziziphus* sp. (Indian jujube) yield fibrous fruits that are consumed during lean seasons. Palms such as *Areca* sp. (betel nut palm) and climbing plants like *Calamus dilaceratus* (Andaman Rattan) and *Korthalsia laciniosa* (Korthalsia palm) offer edible fruits and other nutritional components, demonstrating the macaques' ability to adapt to resource fluctuations. The macaques also forage on species like *Terminalia porocera* (Andaman terminalia) and *Dinochloa andamanica* (Andaman bamboo), consuming leaves, seeds, and stalks as fallback foods. These species play a critical role during resource-scarce periods, highlighting the macaques' ability to adjust their diet to seasonal availability.

Additionally, *Casuarina equisetifolia* (Australian pine or ironwood) plays a supplementary but ecologically important role in the foraging strategies of Nicobar long-tailed macaques, particularly in coastal habitats. Although not a primary food source, macaques exploit various parts of the tree – leaves, seeds, and occasionally young fruits, especially during periods of low fruit availability (Mazumder & Kaburu, 2020). The fine, needle-like leaves are consumed during lean seasons, while seeds are accessed through behaviors such as branch shaking or manipulation to release them from their pods. The presence of *C. equisetifolia* along coastal zones provides an alternative resource when other plant foods or intertidal resources are scarce. The use of this species highlights the macaques' dietary flexibility and their capacity to adjust foraging strategies in response to seasonal fluctuations in resource availability, with potential implications for seed dispersal and ecosystem functioning.

This broad dietary repertoire not only ensures their survival in the dynamic environments of the Nicobar Islands but also underscores their ecological importance. Additionally, this diverse plant utilization underscores the macaques' dietary plasticity and ecological adaptability (Luncz et al., 2017). Their feeding behaviors ensure their survival and contribute significantly to forest regeneration by dispersing seeds of these plant species. This vital ecological service emphasizes the symbiotic relationship between the macaques and their environment, highlighting the need to conserve these habitats and their botanical diversity to safeguard the species.

Additionally, their dietary habits have profound implications for the islands' ecosystem. As frugivores and seed dispersers, they play a pivotal role in forest regeneration, influencing plant community dynamics and biodiversity (Lucas & Corlett, 1998). Their consumption of leaves and flowers affects vegetation structure and growth patterns, while their predation on insects contributes to the regulation of insect populations.

The macaques' dietary flexibility also underscores the importance of conserving the diverse habitats of the Nicobar Islands. The availability of fruits, leaves, and insects is closely tied to the health of the islands' ecosystems, and any disruption to these habitats – whether through deforestation, climate change, or human encroachment – could have cascading effects on the macaques' survival and the broader ecological balance.

The Nicobar long-tailed macaque exemplifies the adaptability and resilience required to thrive in a dynamic and resource-variable environment. By alternating between frugivory, folivory, and insectivory, these primates demonstrate an unparalleled ability to exploit the full spectrum of available resources. Their dietary choices are not only a testament to their ecological versatility but also a critical factor in maintaining the health and stability of their habitat. Understanding the dietary adaptations of these macaques provides valuable insights into their ecological role and the conservation strategies needed to protect their unique insular ecosystem. Ensuring the preservation of their diverse food sources and habitats is essential for safeguarding the future of this remarkable species and the delicate ecological balance of the Nicobar Islands.

3.4 The Scavengers and Tool-Bearers

The Nicobar long-tailed macaque represents a fascinating example of behavioral adaptability in primates, exhibiting a combination of scavenging and rudimentary tool use (Mazumder, 2022). These behaviors underscore the species' remarkable ability to exploit its environment, even amid ecological challenges and anthropogenic pressures. This section explores the macaques' scavenging habits, their limited and context-dependent object use, and the ecological and evolutionary implications of these behaviors.

3.4.1 Scavenging – Opportunistic Feeding Strategies

Scavenging is a prominent feature of the Nicobar long-tailed macaque's feeding behavior, particularly in areas where their habitats intersect with human settlements. Discarded food items such as coconut husks, fruit scraps, and even cooked leftovers offer an easily accessible source of nutrition (Mazumder & Kaburu, 2020). In coastal regions, monkeys other than the Nicobar long-tailed macaques have been observed scavenging for fish (Casey, 2008; Stewart et al., 2008), often competing with other scavengers such as birds or feral dogs (Riley et al., 2015).

This opportunistic feeding strategy reflects the macaques' ability to adapt to anthropogenic changes in their environment. Human activities such as agriculture and coconut farming provide a steady influx of supplemental resources, reducing the energy expenditure required for foraging in the wild. However, this increased reliance on human-provided food also brings risks, including heightened human–macaque conflict (Das et al., 2020). Scavenging macaques frequently raid agricultural fields, homes, or trash sites, negatively impacting local communities (Hasan et al., 2023; Rifaie et al., 2024). Such conflicts underscore the need for strategies that balance macaque conservation with human livelihoods.

Despite these challenges, scavenging highlights the macaques' resourcefulness. By capitalizing on anthropogenic resources, scavenging may increase during periods of reduced fruit availability, although quantitative data linking scavenging frequency to seasonal scarcity in Nicobar macaques remain limited. Future studies comparing scavenging behavior across seasons and across sites with varying degrees of human presence would help clarify this relationship.

3.4.2 Tool Use and Object Manipulation in Ecological Context

Besides scavenging, Nicobar long-tailed macaques exhibit rudimentary tool use and/or object manipulation, a rare and significant behavior among primates (Mazumder & Kaburu, 2020; Pal et al., 2017). In this section, a *tool* is defined as an external object, which may be an inanimate item, or under specific conditions another individual, that is actively held or manipulated by an organism (the user) to simplify or enhance the efficiency of a process, conserve energy (conditionally), and accomplish specific tasks such as foraging, protection, hygiene, or social interaction. Tools may, but do not necessarily, involve intentional morphological or functional modification and reflect adaptive utility or the exploitation of specific affordances within the user's ecological or social context (Mazumder & Chauhan, 2025). Under this definition, only object manipulations with a clear functional outcome are treated as tool use, whereas exploratory or play-related handling of objects is classified as object manipulation. Observed behaviors include:

- **Use of Stone or Wooden Substrates for Food Processing:** Nicobar long-tailed macaques have not been observed using handheld stone hammers; rather, they selectively use stable substrates to assist in processing hard-shelled food items.
- **Stick Manipulation (Object Manipulation):** Juveniles and adults frequently manipulate sticks and twigs in exploratory or play contexts. Under the definition used here, these behaviors are considered object manipulation unless a clear functional outcome is demonstrated.

- **Leaf Manipulation:** Using leaves as grip aids or intermediaries when handling food that may otherwise be difficult to grasp, such as items with thorns or splinters on their surface (Figure 3.6).

These behaviors are consistent with basic repeated motor adjustment and sensitivity to object affordances. For example, pounding coconuts requires physical strength and the ability to select an appropriate substrate and adjust the force applied to achieve the desired result without destroying the edible interior. Similarly, stick play manipulation may contribute to motor skill development, although its primary function may be exploratory or social. Furthermore, stick manipulation was observed primarily in solitary or non-social contexts and appears to function as exploratory object handling rather than functional tool use. Such behaviors may contribute to familiarity with object properties and occur outside direct foraging activities (Bjorklund & Gardiner, 2010). The prevalence and complexity of tool use among long-tailed macaques (in India and outside) vary across individuals and groups, influenced by environmental conditions, resource

Figure 3.6 (A and B) A Nicobar Long-Tailed Macaque (*Macaca fascicularis umbrosus*) Uses Leaves as Makeshift Gloves to Protect Its Hands from Sharp Coconut Splinters.

availability, and opportunities for social adaptation. Findings from studies on interaction-mediated tool use suggest that such behaviors may support behavioral variation and development within social contexts (Viaro et al., 2023). Through observation and interactive processes, macaques integrate ecological challenges with social dynamics, highlighting how object-use behaviors vary with ecological and social conditions in primates. For example, macaque groups residing in areas with abundant coconuts do not necessarily demonstrate more frequent or advanced coconut-pounding behaviors compared to groups in regions where such resources are limited (Mazumder & Kaburu, 2021). This observation suggests that populations experiencing resource scarcity exhibit greater adaptability, optimizing their behavior to maximize the utility of the resources available to them.

3.4.3 Behavioral Transmission and Learning

Interpretations of tool-use learning in Nicobar long-tailed macaques are informed primarily by comparative studies on other populations of *Macaca fascicularis*, particularly long-tailed macaques in Thailand and Southeast Asia, where tool use during foraging has been documented in greater detail (Luncz et al., 2017; Tan et al., 2018; Tan et al., 2018). In the Nicobar population, direct evidence for social learning remains limited, and inferences are therefore drawn cautiously from observed behavioral patterns and comparative primate literature. The use of tools by Nicobar long-tailed macaques may involve a combination of individual exploration and socially mediated learning processes, as suggested by comparative studies of long-tailed macaques in other regions. Juveniles may play an important role in the persistence of tool-use behaviors through repeated observation and exposure to skilled group members. They often observe skilled group members during foraging activities, with repeated exposure potentially facilitating the gradual refinement of their own actions (Mazumder & Kaburu, 2021; Tan et al., 2018). Repeated observational exposure may be associated with the persistence of tool-use behaviors within macaque populations; however, demonstrating social learning conclusively would require network-based or diffusion analyses, which are beyond the scope of the present dataset. Social tolerance within groups further supports the dissemination of tool-use knowledge. For example, dominant individuals often allow subordinates or juveniles to observe their foraging and processing activities, providing opportunities for skill acquisition (Mazumder & Kaburu, 2021; Tan, 2017). Additionally, the sharing of processed food, such as coconut fragments or termites, reinforces group cohesion and fosters cooperative interactions that facilitate learning (Mazumder & Kaburu, 2021).

3.4.4 Ecological and Evolutionary Implications

The dual behaviors of scavenging and tool use have significant ecological implications for the long-tailed macaque. By exploiting anthropogenic resources, the macaques act as secondary consumers in human-altered ecosystems, influencing nutrient cycling and waste decomposition (Sha & Hanya, 2013b). Their tool use, meanwhile, reflects their adaptability to resource scarcity and their ability to modify their environment to meet their needs (Mazumder & Kaburu, 2020).

From an evolutionary perspective, the macaques' tool use offers valuable insights into ecologically grounded variation in behavioral complexity in primates. While tool use is more commonly associated with great apes, the presence of such behaviors in Nicobar long-tailed macaques suggests that the evolutionary pathways leading to tool use may have been more widespread and varied than previously thought. Comparative studies between the Nicobar macaques and other tool-using species, such as chimpanzees and capuchins, highlight both similarities and differences in the contexts and mechanisms of tool use. For instance, whereas chimpanzees often use tools for hunting or for social displays, macaques primarily use tools for foraging, reflecting their unique ecological pressures and opportunities (Vauclair & Anderson, 1994).

The Nicobar long-tailed macaque exemplifies the interplay between ecological challenges and context-dependent behavioral flexibility. These primates demonstrate a remarkable ability to adapt to diverse and dynamic environments through scavenging and rudimentary tool use. These behaviors enhance their survival prospects and offer a comparative insight into ecological and behavioral pathways associated with object use in primates. As both scavengers and tool-bearers, the Nicobar macaques serve as a living model for understanding the adaptive potential of primates in the face of environmental change. Protecting their habitats and mitigating human–macaque conflicts are critical steps in ensuring the survival of this unique species and preserving its contributions to the ecological and scientific understanding of primate behavior.

3.5 Macaques and Behavioral Plasticity – Food Sharing in a Wild Macaque Population

The Nicobar long-tailed macaque exhibits patterns of food sharing that have been infrequently documented among macaques compared to apes. This pattern, which remains uncommon among documented macaque populations within the family Cercopithecidae (monkeys of Africa and Asia), highlights the behavioral plasticity of this insular subspecies (Mazumder & Kaburu, 2021). Food sharing among Nicobar macaques occurs in varied contexts and serves multiple ecological and social functions, underscoring its critical role in their survival and social organization.

3.5.1 Characteristics of Food Sharing in Nicobar Macaques

Food sharing in Nicobar long-tailed macaques primarily involves the transfer of food from one individual to another, either actively or passively (Mazumder & Kaburu, 2021). Active sharing occurs when the food possessor deliberately offers portions of their food to another individual, often through physical gestures or tolerance during feeding. Passive sharing, on the other hand, involves individuals consuming food discarded or left unguarded by the possessor.

While food sharing in many primates is primarily restricted to mother–offspring contexts, Nicobar macaques have also been observed sharing food between unrelated individuals and across age and sex classes (Mazumder & Kaburu, 2021). Observations have documented instances of males sharing food with other males, juveniles receiving food from adults, and even dominant individuals tolerating subordinates feeding on their leftovers. Such behaviors suggest relatively high levels of social tolerance compared to many other macaque populations, which are typically characterized by competitive and hierarchical social structures (Brown et al., 1991; Feistner & McGrew, 1989a; Hohmann, 2009; Jaeggi & Gurven, 2013; Jaeggi & Van Schaik, 2011). The practice of food sharing among Nicobar long-tailed macaques serves several important functions:

- **Strengthening Social Bonds**: Food sharing is often accompanied by affiliative behaviors such as grooming, vocal exchanges, and physical proximity, all of which reinforce social bonds within the group (Goldstone et al., 2016; Samuni et al., 2018). Males frequently share food with other males, which may strengthen coalitions and alliances essential for maintaining dominance and access to resources.
- **Conflict Mitigation**: By tolerating food sharing, dominant individuals may reduce aggressive interactions and maintain group harmony (Jaeggi et al., 2010). Sharing food can serve as a mechanism to appease subordinates or prevent harassment, thereby minimizing energy expenditure on conflict and maximizing the efficiency of resource use.
- **Nutritional Support**: Food sharing ensures that all group members, particularly juveniles, lactating females, and other dependents, receive adequate nutrition (MacKinnon & Fuentes, 2011). In the unpredictable environments of the Nicobar Islands, where resource availability can fluctuate seasonally, sharing food enhances the survival prospects of vulnerable individuals, contributing to overall group stability.

- **Exposure-based Learning and Familiarization:** Juveniles often receive food from adults, allowing them to learn about new food items and foraging techniques (Blurton Jones, 1987). Exposure to food sharing may facilitate juveniles' familiarity with food items and foraging contexts, supporting skill acquisition through repeated exposure and proximity to experienced individuals.

3.5.2 Ecological and Evolutionary Basis

The ecological challenges of insular living provide a plausible basis for the evolution of food sharing in Nicobar long-tailed macaques. Insular environments, such as those of the Nicobar Islands, are often characterized by limited and patchy resource availability, necessitating cooperative strategies to ensure group survival.

Food sharing may have emerged as an adaptive response to these ecological pressures, enabling macaques to buffer against resource unpredictability (Mazumder & Kaburu, 2021). For instance, sharing coconut fragments or termite-laden wood, both high-value and monopolizable resources, ensures that the benefits of these energy-dense food items are distributed among group members, enhancing overall fitness.

From an evolutionary perspective, food sharing in Nicobar macaques represents a convergence of ecological necessity and social adaptation. While kin selection may account for sharing between mothers and offspring, the occurrence of sharing among unrelated individuals suggests the possibility of reciprocal benefits, such as tolerance, grooming, or coalitionary support (Schaub, 1996). The occurrence of sharing among unrelated individuals is consistent with hypotheses invoking reciprocal benefits, such as future tolerance, grooming, or coalitionary support, or access to resources controlled by the recipient.

3.5.3 Comparisons with Other Primates

Food sharing has been extensively documented in great apes, such as chimpanzees and bonobos, where it is often linked to cooperative hunting, coalition building, or sexual strategies (de Waal, 1989; Feistner & McGrew, 1989b; Fruth & Hohmann, 2018; Goldstone et al., 2016). In contrast, food sharing in Cercopithecidae has been less frequently documented than in apes, making the Nicobar macaques a noteworthy case within current records.

In great apes, food sharing often involves high-value items such as meat or cultivated fruits, whereas in Nicobar macaques, sharing includes a broader range of resources, from coconuts to termite logs. Additionally, while harassment is a common driver of food sharing in apes, Nicobar macaques display a more tolerant approach, with food possessors often willingly sharing without aggressive solicitation (Mazumder & Kaburu, 2021).

This contrast highlights the particular social and ecological pressures faced by Nicobar macaques and suggests that their food-sharing behavior may provide insights into the early evolution of cooperative behaviors in primates.

3.5.4 Behavioral Plasticity and Adaptation

The food-sharing behavior of Nicobar macaques illustrates their context-dependent behavioral adjustment. This adaptability is crucial for navigating the challenges of their insular environment, where resource scarcity and competition demand context-dependent and cooperative strategies.

Their adaptability is further evident in the variability of food-sharing contexts and recipients. For example, food sharing is more common in groups with high resource monopolization, such as those with abundant coconut palms, where only a few individuals possess the skills to access the resource (Mazumder & Kaburu, 2021). In these scenarios, food sharing mitigates the risk of conflict and ensures equitable distribution within the group.

Moreover, the social dynamics of food sharing are shaped by group composition and hierarchy. In larger groups, where competition is higher, food sharing often involves dominant individuals and their allies, reinforcing existing social structures. In smaller or more egalitarian groups, sharing is more evenly distributed, reflecting a cooperative approach to resource management.

The practice of food sharing within Nicobar long-tailed macaques represents an uncommon behavior among documented macaque populations, highlighting the interplay between ecological pressures and social complexity. Their ability to share food, despite competition and hierarchical structures, reflects social tolerance and rank-appropriate interactions within the group. This behavior underscores their behavioral plasticity and provides a window into the evolutionary pathways of cooperation and altruism in primates (de Waal, 2000; Jaeggi & Van Schaik, 2011).

Recent studies have further expanded our understanding of macaque social behavior, highlighting how dominance rank influences other opportunistic behaviors, such as 'robbing and bartering' (Peterson et al., 2022). Research conducted on subadult male long-tailed macaques at Uluwatu, Bali, reveals that these macaques exhibit a unique form of resource exchange, where lower-ranking individuals engage in 'robbing' or 'bartering' items from humans in exchange for food. The study by Peterson et al. (2022) suggests that cohort dominance rank significantly affects the success of these behaviors. Higher-ranking individuals are more likely to dominate interactions and successfully acquire food, while subordinates may rely on alternative strategies, such as collaboration with other macaques or humans, to secure resources.

Further supporting this, the work of Leca et al. (2021) reveals a culturally maintained token economy among free-ranging long-tailed macaques. Their study found that macaques acquire object-robbing and object/food-bartering behaviors through social learning, with these behaviors becoming ingrained in the macaques' culture over time (Leca et al., 2021). The macaques exchange objects for food, and this behavior is passed down through generations, creating a stable social tradition that enhances resource acquisition and maintains group cohesion. Importantly, such patterns do not imply intentional teaching, but rather persistence of group-specific behaviors maintained through repeated exposure and social tolerance. Comparable human–primate exchange behaviors have also been documented in India, though in a different form: Dasgupta et al. (2025) describe how free-ranging Hanuman langurs employ elaborate food-solicitation gestures in urban temple settings to obtain food from humans (Dasgupta et al., 2025). While not identical to the robbing–bartering system observed in Bali macaques, this case underscores the broader spectrum of primate–human exchange dynamics in South Asia.

This ability to adapt behaviorally to both social structures and ecological circumstances highlights the behavioral flexibility of long-tailed macaques. In food-sharing situations, individuals with higher social status can dominate access to food, but their tolerance and the cooperative strategies employed by lower-ranking individuals maintain group cohesion. The plasticity seen in these macaques' behavioral strategies – whether in foraging, food sharing, or engaging in 'robbing and bartering' – reflects their social flexibility and capacity to adjust behavior to social and ecological contexts. The incorporation of social dynamics into food-sharing strategies emphasizes the role of behavioral flexibility in the macaques' survival, allowing them to adapt in response to changing environmental and social conditions.

3.6 Object Manipulation and Food Processing Tools

The Nicobar long-tailed macaque demonstrates remarkable proficiency in object manipulation and tool use, behaviors that are noteworthy for their rarity among primates outside the great apes. These abilities are primarily associated with foraging and food processing, but their applications extend to other behavioral contexts, such as hygiene, play, and communication (Mazumder & Kaburu, 2020; Pal et al., 2017). The organization and contextual deployment of these behaviors underscore the macaques' behavioral flexibility in response to environmental challenges of the Nicobar Islands.

Nicobar long-tailed macaques utilize objects across several behavioral contexts, with foraging being the most prominent. Research has categorized their object use into six key domains: foraging, hygiene, self-directed behavior, play, communication, and environmental exploration (Mazumder & Kaburu, 2020). These behaviors reflect behavioral flexibility and context-dependent object use required to exploit available resources effectively in their insular habitat. The following sections focus primarily on foraging-related object use, with other contexts discussed where relevant.

3.6.1 Foraging and Tool Use

In their foraging activities, Nicobar macaques display a range of tool-use behaviors tailored to the diverse resources in their environment:

- **Stone Tools:** While some macaque populations are known to use stones to crack open hard-shelled nuts, fruits, and mollusks – selecting tools based on size, weight, and shape, which reflects an understanding of the physical properties required for effective tool use (Gumert et al., 2009; Gumert & Malaivijitnond, 2013; Haslam, 2018; Haslam et al., 2013; Malaivijitnond et al., 2007). Nicobar long-tailed macaques have not yet been documented using stones as hammers. However, they have been observed to selectively use stone or wooden platforms as stable substrates, rather than portable anvils, for processing hard-shelled food items such as coconuts. This behavior suggests a distinct form of object manipulation and environmental selection for food processing in this isolated, insular population. Such behaviors were rare and opportunistic and should not be interpreted as routine foraging tool use comparable to stone-hammering populations of long-tailed macaques elsewhere.
- **Leaf Manipulation:** Leaves are used as versatile tools, serving as grip pads to handle sticky or slippery food items or as makeshift containers for collecting smaller prey, such as termites or water (Hockings et al., 2015; Laínez, 2016; Mazumder & Kaburu, 2020). These applications minimize waste and demonstrate the macaques' resourcefulness in utilizing natural materials and object manipulation skills. Macaques utilize leaves to handle food that may otherwise be difficult to grasp, such as items with thorns or splinters on their surface. This behavior not only reduces the risk of injury but also enhances the macaques' ability to manipulate and consume otherwise inaccessible resources. Similar behaviors have been observed in other primate species, indicating a broader trend of tools and materials used to optimize foraging efficiency (Beck, 1980; Shumaker et al., 2011). This behavior reflects context-dependent object

use and motor adjustment to material constraints, without implying innovation, planning, or abstract problem-solving.

- **Stick Use:** Nicobar long-tailed macaques have been observed using sticks not only for foraging but also for refining tools through a process of scraping (Mazumder & Kaburu, 2020). In some instances, macaques will scrape or shave twigs and branches to modify their shape, making them more effective for other uses, such as foraging or object manipulation. This behavior involves repeated manipulation of material properties and motor refinement, without evidence for foresight, planning, or symbolic understanding. By scraping the surface of a twig or branch, macaques reduce its roughness, creating a finer, more precise tool that can be used to extract insects and larvae or for self-stimulation. This refining behavior demonstrates the macaques' ability to adapt natural materials to better suit specific tasks, illustrating their adaptability and resourcefulness. The act of modifying a stick to make it finer for a particular purpose reflects a level of behavioral flexibility through repeated interaction with material properties that increase functional effectiveness. The use of sticks in this way not only showcases the macaques' tool-use behaviors but also highlights their ability to adapt and modify available resources for different environmental challenges. This flexibility further emphasizes the macaques' behavioral plasticity, as they can adapt their use of objects to meet various ecological and social demands.

  Nicobar long-tailed macaques exhibit a variety of behaviors involving stick use, particularly in non-foraging contexts (Mazumder & Kaburu, 2020). Juveniles frequently utilize sticks in playful interactions, engaging in exploratory or social activities that involve manipulation of these objects. Such behaviors play a crucial role in the development of motor skills and social bonds. Additionally, stick play serves as a foundation for building skills, allowing individuals to refine techniques that may later be applied in practical scenarios (Albanese et al., 2022; Ballesta et al., 2014). The dual role of sticks in both recreational and skill-building activities highlights the macaques' adaptability and context-specific approach to object manipulation.

  In addition to their foraging and play behaviors, Nicobar long-tailed macaques also use sticks for self-stimulation and sensory engagement (Mazumder & Kaburu, 2020). These macaques, particularly adult males, have been observed using sticks as a form of tactile stimulation, often manipulating or holding them in ways that appear to provide sensory feedback. This behavior is particularly evident when macaques interact with sticks during idle moments or as part of exploratory behavior. The use of sticks for stimulation could serve several purposes. For one, it may fulfill a sensory or exploratory need, where the macaques engage with objects in their environment to gain tactile or proprioceptive stimulation. This behavior may also serve as a coping mechanism during periods of stress or boredom, helping to soothe or distract the macaques when other activities or social interactions are not immediately available. Such behaviors reflect flexible object engagement and sensory exploration, relying on repeated interaction with materials (such as sticks, for a variety of purposes, ranging from functional foraging and play to more subtle forms of sensory stimulation and self-soothing) rather than foresight or goal-directed planning. This flexibility in object use underscores the macaques' behavioral plasticity and their ability to adapt to the challenges and opportunities within their dynamic environments.

- **Bird Feathers or Nylon Threads:** Nicobar long-tailed macaques exhibit innovative grooming behaviors, including the use of bird feathers and nylon threads for flossing (Mazumder & Kaburu, 2020; Pal et al., 2017). In this behavior, macaques collect bird feathers or discarded nylon threads from their environment and use them to clean between their teeth, effectively removing food particles or debris. The fine, flexible nature of both bird feathers and nylon threads allows the macaques to manipulate these materials with precision, performing a flossing-like action that contributes to their oral hygiene. This behavior demonstrates the macaques' flexibility and resourcefulness, as they adapt objects in their surroundings for specific, health-related purposes. The use of bird feathers for flossing may reflect a natural inclination to keep their teeth clean, while the incorporation of nylon threads shows an adaptation to anthropogenic resources, highlighting the macaques' ability to use human-made materials for practical purposes. Such behaviors underscore the macaques' behavioral plasticity, as they repurpose materials from their environment in new and functional ways. This behavior reflects their adaptability and manipulation of material properties without implying foresight or planning, allowing them to engage in self-care practices that enhance their survival. The integration of both natural and artificial materials for flossing illustrates the macaques' adaptability in responding to their ever-changing ecological surroundings.

The tool-use repertoire of Nicobar long-tailed macaques extends beyond direct food acquisition and includes a range of object-manipulation behaviors that reflect behavioral flexibility rather than task specialization. In this population, sticks are used predominantly in play and self-directed contexts. Evidence from closely related long-tailed macaque populations, where tools are more regularly employed in extractive foraging, provides important comparative insight into how similar cognitive capacities may be expressed differently under varying ecological and material conditions.

One notable example is tool-assisted water scooping and drinking, as documented in Balinese long-tailed macaques (*Macaca fascicularis*) in Ubud, Indonesia (Cenni et al., 2023). Over a four-year study, individuals were observed repeatedly using leaves, stones, nuts, and pits as tools to scoop and drink water. This behavior, while functional, was also performed in playful contexts, underscoring the overlap between instrumental and non-instrumental object use in macaque populations. The habitual use of tools for water-related tasks in this population is associated with their broader engagement in object manipulation, including a cultural practice of stone handling. This unique behavior suggests a continuum of tool-assisted and playful actions, reflecting both environmental adaptations and persistence of group-specific behaviors (pending formal tests of transmission).

Although such behaviors have not been documented in Nicobar long-tailed macaques to date, their proclivity for tool use, object manipulation, and exploratory behaviors suggests that similar practices could emerge under appropriate ecological conditions. The inclusion of tool-assisted drinking behaviors expands the scope of tool-use research, offering valuable insights into the relationship between necessity, play, and cultural tradition in primates. By incorporating these observations, future studies can better explore the cognitive and cultural dimensions of tool use across macaque populations.

3.6.2 Trial-and-Error Method and Behavioral Flexibility

The tool-use behaviors of Nicobar macaques highlight their adaptability. For instance, when processing hard-shelled fruits, individuals often employ trial-and-error strategies to identify the most efficient methods. This may include:

- Biting the shell to create an initial crack.
- Striking the shell against a hard surface to weaken its structure.
- Utilizing stones or wooden platforms as substrates to complete the task.

Such behaviors are consistent with trial-and-error learning and adjustment based on immediate feedback, without implying abstract understanding or foresight. The flexibility in their tool use enables the macaques to access diverse resources, an adaptation critical for survival in the resource-scarce habitats of the Nicobar Islands.

3.6.3 Social Adaptation and Group-Specific Behaviors

The acquisition of tool-use behaviors in Nicobar macaques may be influenced by socially mediated processes, such as observation, repeated exposure, and tolerance within groups. Juveniles are frequently observed in close proximity to adults during foraging activities, providing opportunities for observation and repeated exposure to object-use behaviors. Importantly, these observations do not constitute evidence of social learning in the strict sense. Demonstrating social learning requires network-based or diffusion analyses that can disentangle social transmission from individual learning or ecological exposure. Such analyses were beyond the scope of the present study. Consequently, references to socially mediated learning here are intended as cautious interpretations consistent with current debates in primate, avian, and cetacean cultural research (Aplin et al., 2012; Hobaiter et al., 2014).

The variation in tool-use behaviors among different groups suggests the possibility of group-specific behavioral traditions, pending formal tests of social transmission. Specific techniques, such as substrate-assisted pounding stone tools or probing for insects, are often unique to particular groups, consistent with the persistence of group-specific behaviors, pending formal tests of transmission (Biro et al., 2003; Haslam et al., 2018). Such patterns highlight the role of social tolerance and repeated exposure in shaping the persistence of these behaviors, without implying formal cultural transmission or cumulative change.

3.6.4 Ecological and Evolutionary Implications

The tool-use and object-manipulation behaviors of Nicobar long-tailed macaques offer insights into ecological adaptation and variation in macaque foraging strategies (Mazumder, 2022). Comparative studies of long-tailed macaques demonstrate substantial variation in tool-use repertoires, ranging from populations that regularly use handheld stone tools during foraging to others that rely primarily on anvils or substrates. The Nicobar macaques appear to occupy an intermediate position within this continuum, relying on stable platforms rather than portable tools. Understanding why some populations develop more extensive tool-use repertoires while others do not remains an important open question.

- **Resource Exploitation:** Tool use allows the macaques to access food resources that would otherwise be unavailable, such as coconuts or insects concealed beneath tree bark. This ability enhances their dietary breadth and provides a significant advantage in their insular environment, where food availability can vary widely.
- **Energy Efficiency:** The use of tools increases foraging efficiency by reducing the time and effort required to access food. For example, using a stone or wooden substrate to process coconuts is more energy-efficient than attempting to bite through the shell. This efficiency enables the macaques to allocate energy to other vital activities, such as grooming or socializing.

- **Evolutionary Insights:** The presence of tool use in Nicobar macaques offers valuable insights into the evolution of behavioral complexity in primates. While such behaviors are often associated with great apes, their occurrence in this macaque subspecies suggests that the ecological and cognitive conditions driving tool use may be more widespread among primates than previously recognized.

The Nicobar long-tailed macaque illustrates the intersection of ecological adaptation and behavioral flexibility through its patterns of object manipulation and limited tool-related behaviors. By using stones, sticks, and leaves in context-dependent ways, these primates can cope with environmental challenges and access a wider range of dietary resources. Such behaviors appear to be shaped by individual experience and social context, underscoring the relevance of group dynamics for the persistence of these practices.

Protecting the habitats and behavioral repertoires of Nicobar macaques is therefore important not only for conserving this insular subspecies but also for maintaining an opportunity to examine how primates respond behaviorally to ecological pressures. These macaques provide a valuable comparative case for understanding the conditions under which object use, tolerance, and flexible foraging strategies emerge, offering insights into broader questions of primate adaptation without presupposing advanced cognition or cultural transmission.

3.7 Conservation and Preservation of the Nicobar Long-Tailed Macaque

The conservation implications outlined here are briefly introduced to contextualize the behaviors discussed in this chapter and are explored in greater depth in Chapter 7. The behaviors of scavenging and tool use exhibited by Nicobar long-tailed macaques demonstrate their adaptive capabilities, which allow them to thrive in the ecologically diverse and sometimes harsh environments of the Nicobar Islands. However, these adaptive behaviors also underscore the growing challenges that the species faces in the face of rapidly changing environmental and anthropogenic pressures. Habitat destruction, human encroachment, and increasing dependence on anthropogenic resources disrupt the balance between their natural and opportunistic foraging strategies, threatening both their survival and the preservation of their unique behavioral repertoire.

3.7.1 Ecological and Social Challenges

The increasing destruction of natural habitats, particularly due to deforestation, agricultural expansion, and infrastructure development, poses a significant threat to the Nicobar long-tailed macaques (Behie et al., 2019; Bridger, 2022; Caro, 2007; Hansen et al., 2024). These activities reduce the availability of essential food resources and alter the macaques' access to vital habitats, such as primary forests and coastal zones. As a result, macaques are increasingly forced to rely on human settlements for food, leading to a disruption in their foraging patterns. Such shifts in resource availability not only affect the macaques' survival but also have broader ecological implications, as these primates play an important role in seed dispersal and maintaining the health of their ecosystems.

Moreover, the increased proximity to human settlements has led to heightened human–wildlife conflict (Karayathil et al., 2023; Riley et al., 2015). The macaques, opportunistic feeders by nature, often scavenge from garbage dumps, crops, and food waste, which exacerbates tensions between local communities and the animals. These interactions frequently lead to negative outcomes for both parties, including crop damage, the spread of disease, and retaliatory actions from humans. Thus, the increasing reliance on anthropogenic resources threatens to undermine the macaques' natural foraging strategies and behaviors, such as food sharing, which are essential for maintaining group cohesion and social dynamics.

3.7.2 *Conservation Strategies*

To effectively conserve Nicobar long-tailed macaques and their unique behaviors, it is essential to adopt conservation strategies that address both ecological and social dimensions (Bezanson & McNamara, 2019). Habitat preservation and restoration are the cornerstones of these efforts. Protecting and rehabilitating forests, coastal areas, and secondary-growth forest habitats that are crucial for macaques' food and foraging activities will help ensure the availability of natural resources that support their food-sharing behaviors and tool use (Beier & Noss, 1998).

In addition to habitat protection, conflict mitigation measures must be prioritized. Community education about the ecological and cultural significance of macaques can foster coexistence between humans and macaques (Dawson et al., 2024). Raising awareness about the macaques' role in ecosystem services, such as seed dispersal and pest control, may reduce negative perceptions and promote more sustainable land-use practices. Additionally, improving waste management systems to reduce human–macaque interactions is critical for minimizing conflict and protecting the macaques' natural behaviors.

Furthermore, behavioral research is essential for understanding the ecological and evolutionary importance of macaques' food-sharing and tool-use behaviors (Choudhury, 1988). Detailed studies of these behaviors, including how they are passed down through social learning and how they are influenced by environmental variables, can inform more targeted conservation strategies. However, comprehensive behavioral studies on the Nicobar long-tailed macaque remain limited. Since this subspecies is geographically isolated and ecologically unique, long-term field investigations are urgently needed to document its behavioral ecology, social learning processes, and adaptability to anthropogenic (Mazumder & Kaburu, 2020; Pal et al., 2017; Velankar et al., 2024). Such research would not only fill critical gaps in our understanding of primate adaptation in island ecosystems but also provide a scientific basis for conservation planning in the fragile Nicobar archipelago.

3.7.3 Implications of Food Sharing for Conservation

The food-sharing behavior of Nicobar long-tailed macaques is not only a fascinating aspect of their social structure but also a crucial element of their survival. As macaques rely on both natural and anthropogenic resources, the preservation of habitats is paramount for maintaining their food-sharing dynamics. Food sharing serves multiple functions, including the enhancement of social bonds, the reduction of conflict, and the support of vulnerable individuals within the group, such as juveniles and lactating females. Disruptions to their natural foraging habits, such as reduced access to wild food sources, can undermine these behaviors, leading to increased competition and potentially fracturing social bonds. Protecting the macaques' habitat and minimizing their dependence on human settlements will help maintain the delicate balance between their natural behaviors and opportunistic foraging strategies.

Human activities, particularly deforestation and agricultural expansion, threaten the macaques' access to these essential resources. These activities not only disrupt food-sharing practices but may also force the macaques to alter their social dynamics and foraging strategies, leading to the potential loss of critical behaviors that support group cohesion (Gumert et al., 2013). The growing dependence on anthropogenic resources may further dilute the macaques' ability to engage in natural foraging behaviors, limiting their access to diverse and nutritionally rich food sources. Therefore, safeguarding their habitats is not just crucial for maintaining food availability but also for preserving their social and ecological behaviors.

3.7.4 Broader Implications for Primate Research

The unique behaviors of Nicobar long-tailed macaques, such as tool use and food sharing, contribute significantly to the broader understanding of ecologically grounded variation in primate behavior and social organization. By documenting and studying these behaviors, researchers can gain valuable insights into the evolutionary pathways that led to the development of cooperation, social learning, and context-dependent behavioral responses in primates (Behie et al., 2019; Carvalho et al., 2022). These behaviors, particularly when examined in the context of insular environments, shed light on the cognitive flexibility and adaptability required for survival in such ecosystems.

Furthermore, the study of these macaques offers a window into the complex relationship between ecological pressures and behavioral adaptation. How these macaques navigate the challenges of their environment through behavioral plasticity, behavioral flexibility, exploratory object use, and social cooperation provides important lessons on the mechanisms that drive primate evolution, especially in isolated environments like the Nicobar Islands.

The conservation of Nicobar long-tailed macaques requires a comprehensive approach that addresses both ecological preservation and human-wildlife conflict. By protecting their natural habitats, restoring degraded areas, and fostering coexistence between humans and macaques, we can ensure the survival of this remarkable subspecies and its unique behavioral traits. The macaques' tool use, scavenging habits, and food-sharing behaviors reflect their behavioral flexibility and adaptability. These behaviors not only enhance their survival in a challenging environment but also offer valuable insights into ecologically grounded variation in primate behavior and social dynamics (Albert, 2013; Alcock, 1972; Andrews et al., 2022; Bjorklund & Gardiner, 2010; Choi, 2009; Jaeggi & Van Schaik, 2011). Preserving these macaques and their habitats is essential for maintaining the rich biodiversity of the Nicobar Islands and for advancing our understanding of primate cognition and evolution. As such, conservation efforts must focus on ensuring the continued availability of natural resources, reducing human–macaque conflicts, and conducting ongoing behavioral research to safeguard the species and its unique ecological roles.

Thus, the macaques' group foraging strategies suggest an intricate balance of cooperation and competition, which allows them to exploit a range of resources across diverse habitats. Their dietary plasticity – switching between frugivory, folivory, and insectivory depending on seasonal availability – demonstrates an ability to exploit available food sources efficiently, ensuring their survival even during periods of resource scarcity (Kappeler & Van Schaik, 2002; Velankar et al., 2024). Furthermore, their use of stable substrates and object-assisted processing of hard-shelled fruits and extract insects reflects trial-and-error object use and context-dependent behavioral flexibility in response to ecological challenges (Shumaker et al., 2011).

Equally significant is the macaques' unique food-sharing behavior, which is rare among macaques and provides an important social function within the group. Food sharing likely facilitates social bonding, reduces intra-group competition, and supports the survival of vulnerable individuals such as juveniles and lactating females (Feistner & McGrew, 1989a; Swedell, 2012). This behavior underscores the critical role of social dynamics in macaques' foraging strategies and overall group cohesion.

In conclusion, the food and foraging behavior of the Nicobar long-tailed macaque reflects a flexible interplay between ecological conditions, social organization, and context-dependent behavioral responses. These macaques demonstrate the ability to navigate the challenges of their insular environment through adjustments in habitat use, group foraging dynamics, and dietary breadth, shaped by resource availability and social structure. Their habitat preferences, group-level foraging patterns, dietary flexibility, and context-dependent object use contribute to their adaptive success without presupposing advanced cognition or formal cultural transmission.

Understanding these behaviors enhances our knowledge of primate ecology and social evolution, offering insight into how primates respond behaviorally to ecological pressures, particularly in isolated environments. This perspective also underscores the importance of conservation efforts to preserve the unique biodiversity of the Nicobar Islands. As the macaques' behaviors are increasingly influenced by anthropogenic pressures, further research is essential to clarify how ecological change shapes foraging strategies, social tolerance, and object-use behaviors, and to inform conservation strategies that protect both habitats and the behavioral diversity they support.

4 Social Behavior, Exploration, and Context-Dependent Behavioral Responses in Long-Tailed Macaques

Long-tailed macaques (*Macaca fascicularis*), including the Nicobar subspecies (*M. fascicularis umbrosus*), are renowned for their widespread distribution and extraordinary adaptability across diverse ecosystems (as explained in Chapter 3). Their complex social structures and experience-dependent behavioral responses are a testament to their behavioral flexibility, enabling them to thrive in natural and anthropogenic environments alike. This chapter explores the intricate interplay between their exploratory tendencies, trial-and-error behavioral adjustment, social play, and communication strategies, emphasizing how these elements sustain group dynamics and individual survival. Here, *behavioral flexibility* refers to short-term, context-dependent behavioral adjustment arising from individual experience and repeated exposure, rather than heritable evolutionary change or culturally transmitted norms; whereas *behavioral plasticity* refers to the capacity underlying those responses.

Juveniles engage in exploratory behaviors that foster skill acquisition, spatial awareness, and social understanding, laying the groundwork for adult roles (Tan, 2016). This flexibility, evident in tool-assisted foraging and context-dependent responses to anthropogenic challenges, highlights their capacity for behavioral plasticity. Social play is pivotal in physical development, motor coordination, and the reinforcement of group bonds, often serving as a precursor to functional behaviors like tool use (Ballesta et al., 2014).

Communication is another cornerstone of macaques' social organization, incorporating tactile gestures, vocalizations, and visual signals to convey information about hierarchy, cooperation, and group cohesion (Berenstain, 1986; Oren et al., 2024; Romanski, 2013; Whitehouse et al., 2024). These behaviors contribute to group cohesion and coordination, though their evolutionary implications remain to be empirically established (Joly et al., 2017; Macellini et al., 2012). Together, they offer profound insights into the social and behavioral complexities of this remarkable species (Overduin-de Vries et al., 2024).

DOI: 10.4324/9781003614579-4

4.1 The Explorers

Exploration is a fundamental behavior exhibited by long-tailed macaques, reflecting their innate curiosity and adaptability to diverse ecological niches. Juveniles engage in exploratory behaviors that foster skill acquisition. Research suggests that young macaques preferentially attend to and interact with aspects of their environment and conspecifics, which may shape individual behavioral trajectories. For instance, studies on the development of stone-tool use in long-tailed macaques in Thailand indicate that juveniles progress from simple object manipulation to proficient tool use over time. This progression involves both individual exploration and socially biased exposure, in which juveniles preferentially attend to experienced individuals without evidence of imitation per se (Tan, 2016). Additionally, the social dynamics observed among juvenile macaques reflect the reproductive strategies and social structures they will encounter as adults. Engaging in social play and interactions during the juvenile period allows macaques to assess potential mates and social partners, which is essential for their future reproductive success (Amici et al., 2019). Understanding these developmental processes is vital for comprehending how juvenile exploratory behaviors contribute to long-tailed macaques' overall adaptability and survival in their natural habitats. This section examines the role of exploration, its associated risks and rewards, and its broader implications for the species' survival in the dynamic ecosystems of the Nicobar Islands.

4.1.1 Role of Exploration

Exploration among long-tailed macaques serves several interconnected purposes, each contributing to the individual's and group's overall fitness. Juveniles are particularly engaged in exploratory behaviors, often manipulating objects, venturing into new areas, and interacting with conspecifics to gather information about their environment (Ostner & Schülke, 2018).

- **Behavioral Development:** Juveniles and subadults in macaque species frequently engage in exploratory behaviors, including repeated interaction with objects (e.g., leaves, food items, small branches) and close spatial association with adults, which allow them to become familiar with edible resources and environmental affordances. Experimental studies in captive macaques indicate that observation can influence task engagement and attention, rather than direct imitation (Boussaoud, 2015), while object–context paradigms demonstrate individuals' capacity to respond to recurring relationships between objects and substrates (Abbaspoor et al., 2023). These findings derive from controlled laboratory contexts and are cited here solely for mechanistic comparison. In the resource-limited and isolated environments characteristic of island macaque populations, individually driven exploration and repeated exposure to environmental affordances are likely to play an especially important role in survival.

- **Understanding Social Hierarchies:** Exploration is not limited to physical spaces but extends to social settings. Juveniles interact with various group members, testing social boundaries and observing hierarchical structures (Overduin-de Vries et al., 2024). This behavior helps them learn the dynamics of dominance and submission, which are essential for navigating social relationships as they mature.
- **Resource Acquisition:** Exploring new terrains often leads to the discovery of previously unknown food resources, such as intertidal zones rich in crustaceans or forest patches with seasonal fruits (Mazumder, 2022; Tan, 2017). Juveniles, by observing adults or engaging independently, develop foraging strategies that are critical for their nutritional needs (Tan et al., 2018).

4.1.2 Risk and Reward in Exploration

While exploration provides substantial benefits, it is inherently accompanied by risks. Juveniles, due to their inexperience, are particularly vulnerable during exploratory activities.

- **Predator Encounters:** Although the Nicobar Islands lack large terrestrial predators, young macaques must remain vigilant against aerial predators such as raptors. Their exploratory behaviors, which often involve venturing away from the safety of the group or mother, increase their exposure to such threats (Chandra & Venkataraman, 2014; Etting et al., 2014; Micheletta et al., 2012; Van Noordwijk & Van Schaik, 1986).
- **Social Conflicts:** In macaques, juveniles exploring social hierarchies occasionally provoke confrontations with dominant individuals. For instance, encroaching on high-ranking individuals during feeding or grooming sessions can lead to aggressive responses (Carne et al., 2011; Chamove et al., 1984). These encounters, while risky, also provide valuable lessons in social negotiation. Research on macaques indicates that impulsivity is related to aggression, particularly in more despotic social structures, highlighting the importance of understanding social cues (Anderson & Chamove, 1985).
- **Behavioral Plasticity:** The ability of macaques to navigate these risks while maximizing rewards underscores their remarkable behavioral plasticity. For instance, juveniles quickly learn to retreat or adopt submissive postures in response to aggression, allowing them to continue exploratory activities without severe consequences (Honess, 2021).

Behavioral plasticity enables organisms to adapt to environmental changes, and in primates, this adaptability is crucial for survival in dynamic habitats (Van Buskirk, 2012).

Despite these insights, robust, long-term longitudinal data on Nicobar long-tailed macaques remain scarce, particularly datasets that track individual juveniles across developmental stages. Such data are essential to rigorously demonstrate how early exploratory behaviors translate into adult competence, risk management strategies, and long-term fitness outcomes in this insular population. Nonetheless, these aspects of juvenile behavior in macaques highlight the balance between the benefits of exploration and the inherent risks, emphasizing the species' adaptability.

4.1.3 Adaptive Significance of Exploration

Exploration plays a pivotal role in the adaptive success of Nicobar long-tailed macaques, particularly in the unique and often challenging habitats of the Nicobar Islands.

- **Adaptation to Dynamic Environments**: The Nicobar Islands' ecosystems are characterized by frequent disturbances, including monsoons, cyclones, and human activity (Murty et al., 2006; Porwal et al., 2012). Exploratory behaviors allow macaques to identify and exploit new resources, such as intertidal zones during low tides or anthropogenic food sources in human settlements. This adaptability ensures their survival in environments where resource availability fluctuates dramatically.
- **Socially Mediated Exposure**: Juveniles' exploratory behaviors often involve repeated exposure to adult activities, which may facilitate familiarity with task contexts without implying imitation or intentional teaching (Gumert et al., 2011; Ren et al., 2024; Tan et al., 2018). As highlighted by Tan et al. (2018), juveniles' exploratory behaviors often involve selectively observing and, through repeated social exposure to conspecifics, skilled individuals, particularly those proficient in specific tasks like tool use. This targeted attention plays a crucial role in facilitating repeated exposure to task contexts within macaque groups, without implying imitation or teaching.
- **Group Cohesion**: Exploratory activities often occur in the context of the group, with juveniles engaging in playful interactions or cooperative foraging (Mayhew et al., 2020). Such behaviors not only provide opportunities for skill development but also play a critical role in reinforcing social bonds and maintaining group cohesion (Huffman, 1996). This cohesion is essential for the survival and functioning of highly social

species like macaques, as it facilitates coordinated group activities and mutual support in challenging environments.

4.1.4 Case Studies and Observational Insights

Field observations of Nicobar long-tailed macaques provide compelling evidence of the importance of exploration in their lives. Juveniles have been noted to spend significant portions of their day interacting with novel objects, from natural materials like sticks and leaves to anthropogenic items such as discarded plastic, clothes, and ropes. Such interactions highlight their curiosity and ability to integrate new elements into their behavioral repertoire.

For example, in coastal areas, young macaques have been observed manipulating wood, stones, and leaves, behaviors that may later translate into tool use for accessing intertidal food sources. Similarly, in forested regions, juveniles frequently climb new trees, testing their structural stability and the availability of fruits, which contributes to their spatial awareness and foraging efficiency.

Exploration is a cornerstone of the Nicobar long-tailed macaques' behavioral repertoire, offering insights into their behavioral capacities, experience-dependent processes, and ecological adaptations. While it exposes individuals, especially juveniles, to risks, the rewards in terms of resource acquisition, social understanding, and skill development far outweigh the dangers. The exploratory tendencies of these macaques underscore their evolutionary success as adaptable and intelligent primates capable of thriving in the dynamic landscapes of the Nicobar Islands. Further research into their exploratory behaviors can deepen our understanding of their ecological roles and inform conservation strategies to protect this unique subspecies.

4.2 Exploratory Object Users

The long-tailed macaque (*Macaca fascicularis*) exhibits marked behavioral flexibility across ecological contexts. In the Nicobar Islands, object use and tool-related behaviors occur within dynamic and resource-variable environments, where repeated exposure and individual exploration shape behavioral responses (Luncz et al., 2017). This section explores the facets of their adaptive behavior, including tool use, object manipulation, and adaptation to anthropogenic environments. This section does not redescribe the tool-use behaviors detailed in Chapters 2 and 3; instead, it focuses on the exploratory and social contexts in which object interaction occurs.

4.2.1 Tool Use and Exploratory Object Interaction

Tool-related object manipulation has been documented in some long-tailed macaque populations, though its expression varies widely across ecological contexts. Observations reveal that these macaques use a variety of natural materials, such as leaves, sticks, and stones, for foraging and problem-solving.

- **Functional Tool Use:** The macaques exhibit an impressive ability to adapt tools for specific purposes. For example:
    a) **Sticks:** Nicobar long-tailed macaques have been observed rarely modifying sticks, primarily in non-foraging contexts, such as self-directed behavior rather than routine food processing. The few documented incidents of foraging show that Nicobar long-tailed macaques use sticks as filing tools to refine and smooth the surface of another stick prior to consumption. This behavior not only facilitates easier handling and ingestion but also serves to eliminate splinters or rough edges that could otherwise pose a risk of injury to the oral cavity (Mazumder & Kaburu, 2020). The careful selection and manipulation of sticks reflect sensitivity to their functional properties.
    b) **Stones:** Utilized as stable substrates for processing hard-shelled food items requiring precision and persistence. This behavior highlights repeated selection of stones whose physical properties permit effective processing (Gumert & Malaivijitnond, 2013).
    c) **Leaves:** Often folded or manipulated to remove sticky substances from food or use them as globes for protection from splinters, illustrating their capacity to repurpose natural materials for immediate needs (Alcock, 1972; Mazumder & Kaburu, 2020).
- **Behavioral Adaptation:** These behaviors are consistent with behavioral adjustments arising from individual exploration and repeated interaction with environmental features. When faced with novel challenges, such as accessing food hidden in difficult locations (like coconut meat) or a threat from a larger individual, macaques have been observed attempting multiple actions, including using different surfaces, pounding, rolling, or manipulating available substrates such as iron rods from buildings, until they succeed (Mazumder & Kaburu, 2020). Such sequences are best interpreted as trial-and-error learning based on individual experience, rather than innovation beyond instinctual behavior.

4.2.2 Persistence of Group-Specific Behaviors

Object-use behaviors, particularly those involving tools, are not confined to individual ingenuity but may persist within macaque groups through repeated exposure and individual exploration (Aplin, 2016; Barrett et al.,

2017; Macellini et al., 2012). Such persistence of behaviors across time may reflect ecological constraints and repeated individual adjustments rather than established cultural transmission.

- **Observational Learning:** As explained above and in Chapter 2, juveniles and subadults often learn by observing experienced individuals. For instance, a juvenile watching an adult use a stone to crack open a shell may attempt similar actions through trial-and-error exploration to replicate the behavior, gradually refining the technique over time.
- **Behavioral Traditions:** In the macaque society, certain foraging techniques, such as manipulating intertidal organisms or opening oysters, are passed down as group-specific traditions (Gunst et al., 2007; Hammer et al., 2023; Huffman et al., 2010). These behaviors are tailored to the ecological contexts of specific regions, reflecting how cultural practices evolve in response to environmental pressures (Luncz et al., 2017).
- **Innovation Clusters:** Groups inhabiting resource-scarce environments often display a higher propensity for behavioral variation, suggesting that environmental stressors drive the development and transmission of novel survival strategies (Sanz et al., 2013).

These findings refer to tool-using populations of *Macaca fascicularis* outside the Nicobar Islands and are included here strictly for comparative context.

4.2.3 Adaptation to Anthropogenic Environments

The proximity of long-tailed macaques to human settlements has provided them with opportunities to adapt their behaviors to anthropogenic contexts (Fuentes et al., 2011). These adaptations underscore their behavioral flexibility and resilience in navigating human-modified landscapes (Gruber et al., 2019).

- **Resonance Rod:** In urban and semi-urban areas, macaques exhibit ingenuity in scavenging from garbage, accessing stored food, and exploiting human-made objects. For instance, macaques have been observed utilizing iron rods from unfinished buildings as communication tools in aggressive contexts. Specifically, they have adapted branch-shaking behavior into rod-shaking, producing louder and more resonant sounds to intimidate opponents. This form of tool use, referred to as 'resonance rod' behavior, represents an adaptive strategy that has emerged in response to anthropogenic influences (Figure 4.1) (Mazumder & Kaburu, 2020). It underscores the species' remarkable capacity to repurpose human-made materials to meet ecological and social challenges, highlighting their behavioral flexibility and adaptability in anthropogenic environments.

Figure 4.1 A Long-Tailed Macaque (*Macaca fascicularis umbrosus*) Uses an Exposed Iron Rod from an Unfinished Building, Referred to Here as a 'Resonance Rod,' to Produce Loud, Reverberating Sounds.

- **Robbing and Bartering:** A fascinating example of innovation in human-dominated environments is the phenomenon of 'robbing and bartering' (Brotcorne et al., 2017; Leca et al., 2021; Peterson et al., 2022). Macaques steal objects from humans, such as bags or glasses, and hold them until offered food as a trade. This behavior requires an understanding of cause-and-effect relationships and the ability to manipulate human responses to their advantage.
- **Tool Use in Urban Settings:** The use of tools by macaques extends to anthropogenic materials, with individuals observed utilizing human hair or nylon rods as makeshift dental floss (Mazumder & Kaburu, 2020; Watanabe et al., 2007). Additionally, macaques have been documented employing roads as surfaces for pounding activities, facilitating access to hard-to-reach items (Das et al., 2020). These behaviors highlight the species' ability to adapt to and repurpose human-made materials for functional purposes, and their capacity to incorporate novel materials into their behavioral repertoire.

4.2.4 *Ecological Implications and Behavioral Variations*

The context-dependent behavioral responses of Nicobar long-tailed macaques have significant ecological and evolutionary implications.

- **Survival in Resource-Sparse Environments:** The Nicobar Islands are characterized by fluctuating resource availability, particularly during monsoons or after cyclones (see Chapter 3). The ability to flexibly exploit environmental affordances, from intertidal organisms to anthropogenic food, allows macaques to access a wider range of resources, ensuring their survival in challenging conditions.

- **Behavioral Plasticity and Evolutionary Fitness:** Behavioral flexibility is a trait that enhances the macaques' fitness by enabling them to adapt to diverse ecological contexts. Over time, repeated behavioral adjustment and persistence may contribute to differences in foraging efficiency and resilience to environmental variability (Henrich, 2009; Wcislo, 1989).
- **Behavioral Persistence:** The repeated appearance of similar behaviors across generations may reflect stable ecological affordances and shared exposure, rather than cumulative cultural evolution (van Schaik & Pradhan, 2003). This phenomenon provides valuable insights into the evolutionary roots of culture and cognition in primates, offering parallels to the early stages of human innovation (Biro et al., 2003).

The Nicobar long-tailed macaque exemplifies the relationship between behavioral flexibility, context-dependent adaptive responses, and survival. Their tool use, persistence, and ability to thrive in anthropogenic environments highlight their behavioral flexibility and ecological resilience (Mazumder, 2022; Whiten, 2013). These behaviors ensure their survival in the Nicobar Islands' dynamic habitats and provide critical insights into the ecological pathways shaping primate behavior. Continued research into their context-dependent behavioral responses can deepen our understanding of the adaptive strategies that underpin primates' success across diverse ecological contexts.

4.3 Social Play and Object Manipulation

Social play and object manipulation are integral to the developmental journey of macaques. These behaviors play a critical role in shaping their physical and social competencies, particularly during the juvenile stage when skill acquisition is most pronounced (van Schaik & Burkart, 2011). It is important to note that much of our current understanding of these processes is derived from studies on other populations and subspecies of long-tailed macaques, and direct evidence from Nicobar long-tailed macaques remains limited. Through play, young macaques develop essential survival skills, strengthen social bonds, and lay the foundation for more complex behaviors, including tool use and foraging strategies (Tan, 2017). This section examines the types of social play, the significance of object manipulation, and the role of behavioral acquisition through playful interactions.

4.3.1 Types of Social Play

Social play is a cornerstone of juvenile macaque behavior, characterized by a variety of activities that serve distinct developmental purposes.

- **Chasing and Wrestling:** Engaging in play behaviors such as chasing and wrestling is crucial for the development of juvenile primates. These high-energy activities enable young macaques to develop motor coordination, agility, and physical strength (Hambali et al., 2012; Rajeshkumar et al., 2014). Through chasing, juvenile monkeys practice evasive maneuvers that can later help them evade predators or navigate complex terrains (Paukner & Suomi, 2008). Conversely, wrestling fosters body awareness and builds the strength necessary for climbing and defending oneself in adulthood (Wright et al., 2018). These activities contribute to the development of motor skills, social relationships, and behavioral abilities, which are essential for survival and reproduction in primate societies (Paukner & Suomi, 2008).
- **Mock-Fighting:** Engaging in mock-fighting behaviors allows juvenile macaques to navigate dominance relationships within a non-threatening context. Exaggerated movements, such as play slapping or overemphasized vocalizations, serve as play signals that indicate non-aggressive intentions, thereby maintaining a playful atmosphere (Iki & Kutsukake, 2023). These interactions are crucial for young macaques to comprehend social hierarchies without inciting actual conflict.
- **Cooperative Play:** Similarly, cooperative play activities, including tug-of-war with sticks or group games involving simple objects, promote collaboration among juveniles (Ballesta et al., 2014; Shimada & Sueur, 2018). Such interactions lay the foundation for teamwork, which is essential for groups' foraging and predator-avoidance strategies.

These play behaviors are integral to the social and behavioral development of juvenile macaques, facilitating the acquisition of skills necessary for adult social structures and survival strategies.

4.3.2 *Object Play and Manipulation*

Object play is particularly prominent among juvenile monkeys, involving interaction with natural materials like sticks, stones, and leaves (Bjorklund & Gardiner, 2010; Pelletier et al., 2017).

- **Developmental Significance:** Engagement in object play among juvenile macaques is instrumental in their developmental trajectory, particularly in enhancing motor skills, hand-eye coordination, and behavioral adjustments. For instance, repetitive manipulation of objects, such as picking up and dropping stones, enables juveniles to master control over an object's weight and movement, laying the groundwork for proficient tool use in later stages of life (Huffman & Quiatt, 1986).
- **Precursor to Tool Use:** Object play often transitions into functional behaviors as juveniles mature (Cenni et al., 2024). A young macaque

experimenting with a stick during play may later refine its use for probing crevices or cracking open shells. Such behaviors highlight the overlap between play and practical applications, demonstrating how early exploration fosters skill acquisition.
- **Creativity and Innovation:** Through object play, juveniles exhibit exploratory problem engagement by repurposing objects (Albanese et al., 2022; Bjorklund & Gardiner, 2010). For instance, leaves might be folded to scoop water during play, a behavior that mirrors adult tool use (Cenni et al., 2023). This engagement with objects reflects behavioral flexibility in object interaction.

Collectively, these behaviors underscore the significance of object play in the behavioral and motor development of juvenile macaques, facilitating the acquisition of essential skills for survival and adaptation.

4.3.3 Experience-Dependent Social Exposure through Play

Play serves as a critical medium for individual development and a platform for repeated social exposure among juvenile macaques.

- **Observational Learning:** Juveniles appear to acquire skills through repeated social exposure to conspecifics, although individual learning likely also plays a role. For example, a young macaque watching an older individual manipulate a tool may attempt similar actions, learning through trial and error (Meunier et al., 2007; Tan et al., 2018).
- **Skill Transmission:** Group play facilitates repeated exposure to group-specific interaction patterns (Nahallage et al., 2016). Techniques for object use, vocalizations, and social gestures are often shared through playful interactions, ensuring that critical survival skills are passed down across generations.
- **Conflict Resolution and Bonding:** Playful interactions reduce tension within groups, fostering positive relationships among individuals. Grooming often follows play sessions, reinforcing bonds and contributing to group cohesion (Borries et al., 1994; Grampp et al., 2019; Henazi & Barrett, 1999).

These facets of play are integral to the socialization process, enabling juveniles to acquire essential skills and behaviors necessary for successful integration into adult social structures.

4.3.4 Broader Implications for Survival

Social play and object manipulation are pivotal in the developmental trajectory of Nicobar long-tailed macaques, significantly influencing their survival and adaptability.

- **Preparation for Adulthood:** The skills honed during play – such as encompassing motor coordination, social negotiation, and flexible behavioral responses – directly transferable to adult challenges such as foraging, mating, and conflict resolution. For instance, object manipulation during juvenile stages has been observed to enhance motor skills essential for tool use in adulthood.
- **Behavioral Plasticity:** The flexible nature of play allows juveniles to experiment with a range of behaviors, enhancing their ability to adapt to changing environments. This adaptability is particularly vital in the Nicobar Islands, where ecological conditions are subject to change due to monsoons and human activities. The species' ability to adjust to various habitat conditions, including captive, semi-wild, and wild environments, underscores their behavioral plasticity (Al Hakim et al., 2022).
- **Emotional Development:** Engagement in play not only fortifies physical and social skills but also contributes to emotional resilience. Through participation in low-risk conflicts and playful problem-solving, juveniles cultivate the confidence necessary to navigate more intricate challenges in later life stages. Social behaviors, including play, are integral to their daily activities, facilitating social bonding and reducing tension within groups (Bernstein, 2005; Velankar et al., 2024).

Figure 4.2 A Nicobar Long-Tailed Macaque (*Macaca fascicularis umbrosus*) Caught Mid-Yawn.

Social play and object manipulation are essential components of the long-tailed macaque's developmental trajectory. These behaviors serve as a bridge between exploration and functional application, equipping young macaques with the skills needed for survival in dynamic environments (Vos et al., 1992). By fostering physical strength, social cohesion, and behavioral flexibility, play ensures that the next generation is well-prepared to navigate the challenges of life in the Nicobar Islands. Further research into these behaviors offers valuable insights into the evolutionary significance of play and its role in the adaptive success of primate species.

4.4 Communication Patterns

Communication in Nicobar long-tailed macaques is consistent with those documented in other macaque species. It encompasses a variety of physical, vocal, and visual signals, all of which contribute to maintaining social cohesion, navigating hierarchical structures, and facilitating group survival. Each communication modality serves distinct purposes while often complementing others to convey nuanced messages. This section provides a descriptive overview of communicative behaviors based on observable interactions and contextual use. The interpretations offered are functional rather than ethological or semiotic, and do not imply symbolic communication or higher-order cognitive processes.

4.4.1 Physical Communication

Physical communication forms the foundation of many social behaviors among Nicobar long-tailed macaques. These tactile interactions play a critical role in maintaining relationships, reducing tension, and reinforcing social hierarchies.

- **Grooming:** Grooming is a primary affiliative behavior in macaques, vital for sustaining group cohesion (Gumert, 2007). Grooming provides immediate health benefits by removing parasites and debris, but its social functions are even more significant. It is used to reinforce alliances, especially between individuals of similar rank or those seeking support in potential conflicts. Subordinate individuals often groom dominants as a display of submission, strengthening hierarchical stability (Mishra et al., 2020).
- **Tactile Signals:** Subtle tactile interactions, such as nudges, gentle touches, or pats, serve as a form of non-verbal communication. These signals often initiate social activities like grooming or play, reaffirm bonds between close associates, or soothe distressed individuals. Physical contact is particularly significant in juvenile and maternal interactions, fostering trust and security within the group.

These communication forms are integral to the social structure of Nicobar long-tailed macaques, facilitating interactions that promote group harmony and stability.

4.4.2 Vocal Communication

The vocal repertoire of Nicobar long-tailed macaques is diverse, reflecting their highly social and cooperative nature. Vocalizations facilitate communication across distances, particularly in dense forests where visual signals may be less effective.

- **Alarm Calls**: Alarm calls are essential to the macaques' defense strategy. These calls vary depending on the type and proximity of a predator, such as an aerial raptor or terrestrial threat. For example, high-pitched, repeated calls often signal an immediate danger, prompting group members to seek cover (Palombit, 1992). The specificity and urgency conveyed in these calls demonstrate their capacity to assess and communicate threats effectively. The island lacks major predators, such as large carnivores; as a result, alarm calls are infrequently observed among the Nicobar long-tailed macaques. Furthermore, humans are generally not perceived as a significant threat by the macaques. In both anthropogenic areas and forested habitats, the macaques maintain a cautious approach, consistently keeping a physical distance from humans without emitting alarm calls. This behavior reflects their adaptive strategy for minimizing potential risks while avoiding unnecessary conflict (Mazumder, 2018).
- **Social Calls**: Vocalizations also play a role in maintaining group cohesion, especially during travel or foraging. Low-frequency calls may signal the group's movement direction, while softer vocal exchanges help mothers locate their offspring in dense habitats (Palombit, 1992). Emotional states, such as contentment or agitation, are also communicated through variations in pitch and tone.
- **Individual Variation**: Vocal patterns can vary between individuals, reflecting their social rank, age, or emotional state (Green, 1981; Hohmann, 1991; Palombit, 1992). Dominant individuals may use deeper, more assertive vocalizations to reinforce their status, while juveniles emit higher-pitched calls associated with play or distress. These variations add an individualistic dimension to vocal communication, enabling nuanced interactions.

Understanding the vocal communication of Nicobar long-tailed macaques provides insight into their social structures and adaptive behaviors, highlighting the importance of vocalizations in primate social systems.

Figure 4.3 A Young Male Macaque Mid-Call, Captured in a Moment of Vocal Expression.

4.4.3 *Visual Communication and Tools for Aggression*

Visual cues are equally critical in the communication repertoire of long-tailed macaques. These signals often convey information about dominance, submission, or intent, with high visibility aiding in immediate interpretation by conspecifics.

- **Facial Expressions:** Facial expressions are among the most direct forms of visual communication (Parr & Heintz, 2009; Whitehouse et al., 2024) (Figures 4.2 and 4.3). For instance, baring teeth or opening the mouth wide is a common threat display used to intimidate rivals or assert dominance. Conversely, lip-smacking or lowered eyelids signal submission or appeasement, helping to diffuse tension within the group (Maestripieri, 1996, 2005; Maestripieri & Wallen, 1997).
- **Postural Displays:** Body posture is another significant visual cue (Cant, 1988; Honess, 2021). Dominant individuals often adopt upright stances with direct stares to assert authority, while subordinates crouch, avert their gaze, or lower their bodies to demonstrate submission. Such displays reduce the need for physical aggression by clearly conveying hierarchical status.
- **Aggression and Tool Use:** Although infrequent, there are documented instances of Nicobar long-tailed macaques employing tools in aggressive

contexts. For example, individuals have been observed using iron rods from unfinished buildings as instruments of intimidation, shaking them to produce loud noises aimed at deterring opponents (Mazumder & Kaburu, 2020). This behavior, referred to as 'resonance rod' use, exemplifies the species' capacity to adapt anthropogenic materials for social communication and highlights their behavioral flexibility in addressing environmental challenges.

These visual communication strategies are essential for the regulation of social interactions among Nicobar long-tailed macaques, enabling them to navigate complex social structures and environmental contexts effectively.

4.4.4 Ecological and Social Implications of Communication

The communication strategies of Nicobar long-tailed macaques are finely tuned to their ecological and social environment. In dense forests, the combination of vocal and visual signals ensures effective communication across varying contexts. Physical interactions, such as grooming, strengthen group cohesion, vital for collective survival in resource-scarce habitats. Furthermore, their ability to adapt tools for social and defensive purposes highlights their adaptability and underscores the evolutionary significance of their communication repertoire.

The multifaceted communication patterns of Nicobar long-tailed macaques reflect their intricate social structures and adaptability. These primates maintain group harmony, navigate hierarchies, and address ecological challenges by integrating physical, vocal, and visual signals. Their context-dependent communication behaviors ensure survival and offer valuable insights into the evolution of complex social behavior in primates. Future research focusing on the interplay between these modalities and their group-level behavioral persistence could further illuminate the behavioral capacities of this fascinating subspecies.

In conclusion, the social and behavioral responses of Nicobar long-tailed macaques reflect their exceptional behavioral flexibility and adaptability, traits essential for survival in the dynamic and often resource-scarce ecosystems of the Nicobar Islands. Juvenile macaques, through their exploratory behaviors, engage actively with their environment, honing crucial skills for resource acquisition and social navigation. These early experiences, such as manipulating objects and experimenting with their surroundings, enhance their adaptability and prepare them for the complexities of adult life.

As adults, Nicobar long-tailed macaques exhibit sophisticated communication strategies, including physical interactions, vocalizations, and visual cues, which highlight their complex social organization. Grooming and tactile gestures are central to maintaining social bonds and alleviating tension, while vocal and visual signals play critical roles in coordinating group activities and resolving conflicts. These communication practices reflect the macaques' ability to maintain structured social hierarchies and ensure group cohesion.

Moreover, the macaques' remarkable adaptability extends into anthropogenic environments, where their flexible use of tools and context-dependent foraging strategies allow them to thrive in human-modified landscapes. Such behaviors offer valuable insights into their social organization, experience-dependent processes, and the evolutionary pathways of primates. Despite these observations, communication and behavioral acquisition through exposure remain insufficiently documented in Nicobar long-tailed macaques, highlighting the need for systematic, hypothesis-driven research. A deeper investigation into these areas could provide vital insights into the development of social complexity in primates. Continued research on their behavioral ecology, with a particular focus on communication and behavioral acquisition through exposure, is essential for deepening our understanding of primate adaptability and behavioral diversification.

5 Macaque Social and Reproductive Behavior

The social and reproductive behavior of Nicobar long-tailed macaque (*Macaca fascicularis umbrosus*) is shaped by complex interactions between dominance hierarchies, resource distribution, and mating strategies. These primates exhibit dynamic social structures, where dominance is established and maintained through both physical strength and social maneuvering. Hierarchical positioning influences access to essential resources such as mates, food, and affiliative bonds, ultimately shaping group stability and cohesion. The interplay between dominant individuals, coalition-building, and resource control determines the distribution of reproductive opportunities, while female choice, kin networks, and social relationships further influence reproductive outcomes. Context-dependent behavioral flexibility, expressed within existing social systems, is evident across multiple domains of behavior. Furthermore, the contrast between social and sexual bonds underscores the nuanced relationships that govern group interactions. Examining these behaviors provides insights into the evolutionary pressures expressed within existing social structures, mate selection, and reproductive success in long-tailed macaques.

5.1 Alphas and Power Play

In Nicobar long-tailed macaques, these general patterns of dominance and coalition-building are expressed within small, insular groups where demographic turnover and ecological constraints further shape rank dynamics. Dominance hierarchies in primates are characterized by male-male competition, coalition formation, and rank fluidity, reflecting broader evolutionary strategies observed in primate social structures (Matsumura, 1999; Wrangham, 1980). Unlike species with rigid linear hierarchies, the dominance structure in long-tailed macaques is dynamic, shaped by agonistic encounters, affiliative networks, and shifts in individual fitness over time (Kononowicz et al., 2025; Sueur et al., 2011). Compared to rhesus macaques, which exhibit stricter dominance structures, long-tailed

DOI: 10.4324/9781003614579-5

macaques demonstrate greater flexibility in rank negotiations (van Noordwijk & van Schaik, 2001). Among males, alpha status is not absolute but contingent upon continuous negotiation. High-ranking individuals maintain their position through a combination of physical assertion, strategic alliance-building, and reciprocal relationships with both subordinate males and influential females (Ballesta et al., 2014; Clutton-Brock & Huchard, 2013; van Noordwijk & van Schaik, 1987). While direct aggression and dominance displays serve as immediate mechanisms for asserting control, long-term stability in rank is often achieved through repeated social interactions, alliance formation, and negotiated relationships (Blersch et al., 2024; de Waal, 2007). Coalitionary support plays a crucial role, as alliances between mid- and low-ranking males can effectively challenge alpha individuals, particularly when the dominant male shows signs of senescence or decreased competitive ability.

Maintaining dominance comes with substantial physiological costs. High-ranking males engage in frequent conflicts and dominance displays, leading to elevated stress levels and injuries. Over time, these factors contribute to immune suppression and reduced life expectancy (Dixson, 2012). These strategies include sneaky copulations, forming affiliative bonds with females to secure future mating opportunities, and relying on temporary alliances to navigate competitive social contexts. The role of social bonds in buffering against hierarchical instability suggests that reproductive success is not exclusively dependent on high rank but also on the ability to navigate complex social landscapes (Schülke et al., 2010). The flexibility of dominance structures in long-tailed macaques underscores the adaptive significance of rank negotiation and coalition dynamics (De Laat et al., 2024; Massen et al., 2011).

Studies have revealed significant insights into the longitudinal patterns of rank acquisition and loss, the impact of demographic changes on hierarchy stability, and the physiological costs of maintaining dominance, particularly in response to ecological pressures and anthropogenic disturbances (van Noordwijk & Van Schaik, 1985). These findings enhance our understanding of the evolutionary mechanisms that shape male reproductive strategies and social cohesion within this subspecies. In long-tailed macaques, dominance hierarchies are dynamic, with males acquiring and losing rank based on age, strength, and social alliances (Mishra et al., 2020; van Noordwijk & van Schaik, 2001). Younger, physically stronger males often challenge older, high-ranking males, leading to rank reversals (Figure 5.1) (Noordwijk & Schaik, 2001). Female rank, on the other hand, is more stable and often inherited maternally, with higher-ranking females maturing earlier and having higher reproductive success than lower-ranking females (van Noordwijk & van Schaik, 1987).

Figure 5.1 An Alpha Male Nicobar Long-Tailed Macaque (*Macaca fascicularis umbrosus*) Performs an Open-Mouth Threat Display in Response to Human Proximity.

The fluidity of male dominance is further influenced by broader demographic shifts, including male immigration, the deaths of older individuals, and the maturation of juveniles, all of which can significantly impact rank stability (Anuar & Ullah, 2011; Matsumura, 1999). While alpha males tend to dominate reproduction, subordinate males often maintain a crucial role in the social structure through strategic alliances, which help to stabilize the hierarchy and mitigate the physiological and social costs of rank loss (Matsumura, 1999). As males age, their competitive ability diminishes, and they may lose rank, particularly when confronted by younger, more robust individuals (De Ruiter et al., 1994). The introduction of new males can further disrupt established dominance relationships, creating additional opportunities for rank challenges and shifts in the social structure (van Noordwijk & van Schaik, 2001). Ultimately, the balance between rank stability and social fluidity plays a critical role in maintaining cohesion within macaque groups.

5.1.1 Multiple Male Sub-Groups and Leaders

The social structure of the long-tailed macaque is distinguished by a dynamic system of multiple male sub-groups that engage in both cooperation and competition (Mishra et al., 2020). Unlike rigid social structures observed in many macaque species, these sub-groups are fluid, with males continuously forming and dissolving alliances and rivalries in response to changing group composition, social dynamics, and environmental pressures (De Laat et al., 2024; Massen et al., 2011; Waal & Ferrari, 2012). Leadership within these sub-groups is not static; dominance can shift over time as a result of factors such as physical strength, age, and the strategic use of aggression or ritualized displays (Blersch et al., 2024; Cowlishaw & Dunbar, 1991).

Within these fluid networks, dominant males typically control access to critical resources, including food and mates (Wooddell et al., 2019). However, their dominance is not assured, and power transitions are common, often occurring through subtle or overt challenges. These shifts are not solely determined by physical prowess but are also influenced by the formation of strategic alliances among males or with influential females (Massen et al., 2011; Overduin-de Vries et al., 2014). Consequently, the hierarchical structure is highly dynamic, with leadership roles subject to constant negotiation and recalibration.

Maintaining dominance comes with substantial physiological costs. High-ranking males, frequently engaged in intense conflicts or displays, experience elevated stress levels and physical injuries, which can result in long-term health consequences such as immune suppression and reduced life expectancy (Girard-Buttoz et al., 2014; Matsumura, 1999). The repeated physical confrontations required to sustain dominance may also deplete energy reserves, thereby limiting the male's ability to invest in other areas such as reproduction. As a result, while dominant males benefit from enhanced access to mates and resources, these advantages may be offset by the physiological toll of maintaining a high rank. On the other hand, subordinate males, though less frequently involved in direct physical contests, face their own form of stress due to limited access to resources and mating opportunities, which can impact their reproductive strategies and overall fitness (van Noordwijk & van Schaik, 2001).

Although male dominance plays a crucial role in the social organization of the long-tailed macaque, reproductive success is not solely determined by rank. Female choice significantly influences mating outcomes, complicating the relationship between rank and reproductive success (Clutton-Brock & Huchard, 2013; van Noordwijk & van Schaik, 1987). Female leadership in these groups tends to be more egalitarian compared to the competitive and hierarchical male structure. Power among females is often distributed more equally, with social bonds, especially those formed through maternal kinship, playing a key role in stabilizing group dynamics (van Noordwijk & van Schaik, 1987). While female competition exists, it is typically less intense than the inter-male competition for dominance. This egalitarian structure contrasts with the more hierarchical male dynamics, where power shifts are driven by physical strength, alliances, and reproductive strategies. The complex interplay between inter-male competition, female mate choice, and social alliances highlights the multifaceted nature of the group's hierarchy. Furthermore, the physiological costs associated with rank acquisition and maintenance underscore the influence of both social strategies and biological constraints on the behavior and reproductive success of individuals within the group (Girard-Buttoz et al., 2014). In contrast to male dominance, which shapes competition and power dynamics, female bonds and matrilineal hierarchies are central to the group's stability and reproductive success, fostering long-term social cohesion.

5.1.2 Female Bonds – Hierarchy and Legacy

In primate societies, female bonds are critical for social cohesion, group stability, and reproductive success. Female macaques typically form strong, lifelong alliances, particularly with maternal kin, which serve as the foundation for social cooperation and mutual support (Bercovitch, 1995; Clutton-Brock & Huchard, 2013; Fürtbauer et al., 2010). These bonds are not only vital for the individuals involved but also have far-reaching implications for the group as a whole, fostering a network of protection and assistance against external threats, such as predation and male aggression (Bercovitch, 1995).

While the female dominance structure is often subtler than the male hierarchy, it remains a crucial feature of macaque social organization. Female macaques typically follow a matrilineal hierarchy, where dominance is inherited along maternal lines. High-ranking females generally enjoy better access to key resources and mating opportunities, which directly enhances their reproductive success (Engelhardt et al., 2005; van Noordwijk & van Schaik, 1987). Moreover, the offspring of high-ranking females are more likely to inherit their mothers' elevated status, ensuring better access to resources and higher survival rates. This matrilineal structure not only influences the individual success of females but also shapes the broader reproductive strategies of the group. The offspring of dominant females often ascend to higher social ranks, perpetuating the legacy of their maternal lineage across generations (Abegg & Thierry, 2002; Anuar & Ullah, 2011; Blersch et al., 2024). Consequently, the social and reproductive strategies of the group are deeply intertwined with the family-based hierarchy, where intergenerational bonds reinforce the power and influence of specific matrilines over time (Anuar & Ullah, 2011). The continuity of these female-driven legacies plays a central role in the long-term stability of the group's social order and reproductive success, with dominant matrilines ensuring the perpetuation of their status through successive generations.

5.2 Food Sharing and Macaque Politics

Building on the discussion of social dynamics and power structures, food sharing in macaque societies illustrates how resource distribution reinforces both cooperative and competitive behaviors within the group. As mentioned in Chapter 3, this section focuses specifically on how food sharing functions within social negotiation, alliance formation, and reproductive strategies (Dyble et al., 2016; Feistner & McGrew, 1989a; Jaeggi & Van Schaik, 2011; Mazumder & Kaburu, 2021). Although macaques frequently share food with close kin or allies, the act of sharing is highly context-dependent and often serves as a strategic tool to reinforce alliances or demonstrate dominance (Feistner & McGrew, 1989b; Mazumder & Kaburu, 2021). Far from being a purely altruistic behavior, food sharing often functions within broader patterns of social exchange, where tolerance and access may be associated with future affiliative benefits (Jaeggi & Van Schaik, 2011).

Food sharing is intricately tied to the political dynamics of macaque groups. Food distribution often reflects and reinforces existing power structures, as dominant individuals typically enjoy priority access to food resources. Subordinate macaques, in contrast, must engage in negotiation and social displays, such as deference, grooming, or other submissive behaviors, including observing higher-ranking individuals during foraging without interfering, in order to maintain access to these resources (Mazumder & Kaburu, 2021). These exchanges reinforce hierarchical relationships, with the sharing of food acting as an important social currency that solidifies the position of higher-ranking individuals while maintaining social stability. Thus, food sharing transcends its biological function and becomes a critical social tool that not only facilitates individuals' survival but also reinforces the group's hierarchical and political structures. This behavior underscores the complex interplay between cooperation, competition, and social strategy, highlighting the role of resource distribution in the negotiation of power and status among macaques. Beyond food sharing, another behavior that illustrates the interplay between social bonds and resourcefulness is hygiene practices.

5.3 Females and Hygiene Tools

Food sharing and dominance interactions, such as grooming, serve as another key mechanism for reinforcing social bonds in macaque societies. Grooming facilitates hygiene maintenance and functions as a form of social currency, strengthening alliances and reducing tension within the group. While all macaques engage in grooming, female long-tailed macaques exhibit a notable degree of behavioral flexibility in their grooming practices, incorporating tools or manipulating objects to enhance hygiene maintenance. While less extensively documented than foraging behaviors, this reflects functional material use arising from repeated interactions with environmental affordances.

Field observations indicate that female long-tailed macaques selectively employ materials such as nylon threads and feathers as grooming aids, using them to remove debris from their teeth (Mazumder & Kaburu, 2020; Pal et al., 2017). The selection of grooming materials is consistent with sensitivity to texture and structural properties, without implying advanced planning or specialized cognitive processes. Furthermore, long-tailed macaques in Thailand have been documented using human hair as dental floss, a novel behavior demonstrating the species' ability to incorporate anthropogenic materials into its behavioral repertoire (Watanabe et al., 2007).

The incorporation of objects into grooming contexts is most frequently observed during self-directed hygiene and maintenance behaviors. While such object use may increase the efficiency of grooming actions, there is currently no evidence that it carries social or hierarchical consequences. Parallels between hygiene-related object use and foraging-related tool use suggest that these behaviors may rely on overlapping motor skills and material affordances, rather than implying shared or specialized cognitive mechanisms. In the Nicobar long-tailed macaques, object use in hygiene contexts is predominantly self-directed and appears to serve functional maintenance needs (Mazumder & Kaburu, 2020; Pal et al., 2017). While grooming more broadly plays an important role in social systems, there is currently no evidence that tool-assisted hygiene directly enhances social bonding. Instead, tool-aided grooming reflects material selectivity and behavioral flexibility expressed within an existing socioecological framework.

5.4 Resourcefulness: Individual Variations in Resource Use within Social Hierarchies

In macaque societies, male dominance often dictates access to resources, reproductive opportunities, and social standing (Blersch et al., 2024). However, female macaques exhibit context-dependent variation in resource use and foraging behavior in response to ecological challenges. For instance, female macaques have been observed using natural materials, such as leaves and twigs, for grooming and hygiene (Mazumder &

Kaburu, 2020). These behaviors reflect repeated trial-and-error interactions occurring within existing social and ecological constraints.

Interestingly, resourcefulness in macaque groups can complement, rather than directly challenge, traditional dominance structures in determining social rank (Mazumder & Kaburu, 2020). Female macaques, for example, often display greater variation in object use frequency during foraging, which enables them to secure food more efficiently and adapt to shifting ecological conditions. This capability is evident in species such as the Burmese long-tailed macaque (*Macaca fascicularis aurea*), where sex differences in stone tool use have been observed. Over a two-week study in Laemson National Park, Thailand, it was found that females used stone tools more frequently than males for opening shelled food, particularly sessile oysters. Interestingly, while females employed smaller tools and were observed chipping open oysters using an axing technique, males tended to use larger stones and were more proficient in processing unattached foods using a pounding technique (Gumert et al., 2011).

Males, on the other hand, performed significantly better at opening unattached food items with large pounding stones or manipulating objects and using tools (Gumert et al., 2011; Mazumder & Kaburu, 2020; Pal et al., 2017) (Figure 5.2). Despite this, both sexes regularly employed a variety of techniques. Females struck food items more frequently during tool-use bouts, often at a faster rate, though no significant difference in the duration of tool-use bouts was detected. Males processed food slightly

Figure 5.2 An Adult Male Nicobar Long-Tailed Macaque (*Macaca fascicularis umbrosus*) Selects a Sturdier Pounding Surface, a Pedestrian Path Made of Granite Stones Set in Cement, to Crack Open a Coconut.

faster within a tool-use bout, though no significant difference was observed in the overall rate of food processing during foraging with tools (Gumert et al., 2011). These findings suggest that although males excel at using larger stones to open unattached food, females' frequent use of axing techniques on sessile oysters highlights the broader spectrum of tool-use strategies employed by both sexes.

As documented in Chapters 3 and 4, object use in Nicobar long-tailed macaques occurs across multiple behavioral contexts, including foraging, hygiene, communication, and self-directed activities (Figures 5.2 and 5.3) (Mazumder & Kaburu, 2020), in the present chapter, these behaviors are considered only where they intersect with social interactions or reproductive contexts, rather than as evidence of generalized tool-use capacity.

Observations on Nicobar long-tailed macaques indicate that males engage more frequently than females in object manipulation; however, current evidence does not suggest that such behaviors alter dominance rank, access to mates, or coalition structure. Instead, object use appears to be expressed within established social systems and reflects individual-level variation in behavior rather than shifts in social organization. Consequently, while object use contributes to understanding behavioral flexibility in this subspecies, it does not, on its own, challenge the central role of dominance hierarchies and social bonds in structuring macaque social and reproductive dynamics.

Figure 5.3 A Nicobar Long-Tailed Macaque (*Macaca fascicularis umbrosus*) Uses a Natural Wooden Surface as a Rolling Substrate to Process Food.

5.5 Mate Preference and Selections

Mate selection in long-tailed macaques is a complex process shaped by both direct and indirect factors. Female macaques may exercise direct mate selection by choosing mates based on observable traits, such as social status, physical attributes, and behaviors indicative of genetic fitness (Haunhorst et al., 2020; Kappeler, 2002). Social rank plays a particularly pivotal role in mate choice, with higher-ranking males typically possessing superior access to resources and offering protection, thereby enhancing the likelihood of offspring survival (Girard-Buttoz et al., 2014a, 2014b). Furthermore, physical characteristics and behaviors that signal health and genetic quality, such as strength, symmetry, and aggression, are also significant factors influencing mate selection (Cassini, 2020; Clutton-Brock, 2007; Cowlishaw & Dunbar, 1991; Kappeler, 2002).

Indirect mate selection emerges through broader social dynamics, wherein female macaques may align with or support specific males in competitive interactions, influencing reproductive opportunities (Haunhorst et al., 2020; Stumpf et al., 2011). Such alliances can affect access to mates and enhance the reproductive success of particular males, especially in situations where competition for dominance or resources occurs (Clutton-Brock, 2007).

In addition to these direct and indirect factors, females may also prioritize males who provide material benefits, such as food provisioning or grooming (Clutton-Brock & Huchard, 2013; Kappeler, 2002). These actions not only support female well-being but also contribute to the reproductive environment, with males offering such resources being perceived as more attractive mates. This form of resource provisioning is often linked to reciprocal cooperation, where males demonstrate an ability to invest in the female and her offspring, further increasing their attractiveness as potential mates (De Ruiter et al., 1994; Hector et al., 1989; Kokko & Jennions, 2008).

In turn, male macaques engage in a variety of competitive displays to attract females, including vocalizations, physical posturing, and aggressive behaviors. These displays serve to signal dominance and genetic fitness, thereby enhancing the male's chances of securing a mate (Milich & Maestripieri, 2016; Mishra et al., 2020; Pal, 2018). Thus, mate selection in long-tailed macaques is governed by a dynamic interplay of direct female preferences, male competition, and material provisioning, underscoring the complexity of reproductive strategies in macaque societies.

5.6 Use of Sex toys

The use of external objects in mating behavior, though infrequent, offers an intriguing perspective on sexual behavior in non-human primates. In addition to the typical use of tools for foraging and hygiene, some primates have been observed employing objects such as sticks, stones, and other natural materials in mating contexts. The function of these objects remains partially understood, but they may serve purposes such as sexual stimulation, sexual play, or enhancing mate attraction (Cenni et al., 2022). This form of behavior represents rare, self-directed sexual interactions involving available environmental materials. One notable example of this behavior is the use of tools as a form of sexual stimulation, where Nicobar long-tailed macaques have been observed using a specific object, referred to as a 'stimulation tool,' during self-directed behaviors. Such tools, including items like sticks, are used to facilitate sexual activities, such as masturbation (Mazumder & Kaburu, 2020). This behavior illustrates context-dependent, self-directed object interaction in sexual contexts, without evidence that it enhances reproductive success or functions as a communicative display.

Some studies have discussed object interaction in sexual contexts without evidence that such behaviors signal mate quality or directly influence reproductive success (Cenni et al., 2022). Although tool use in sexual contexts is not widespread, it may reflect macaques' ability to adjust their mating strategies in response to environmental and social factors, adding an additional layer of complexity to their sexual behavior. There is currently no proven evidence that object use in sexual contexts functions as a display, influences mate choice, or involves communicative or symbolic intent; these behaviors are best interpreted as rare, self-directed interactions with available materials.

The occurrence of these behaviors may reflect a combination of environmental exposure and individual experience. In highly competitive environments where access to mates or resources is limited, the incorporation of objects into mating rituals may serve to increase an individual's chances of reproductive success (Cenni et al., 2022; Manson, 1994). Additionally, social learning may play a role, as younger individuals are frequently exposed to the behaviors of older or more experienced group members, increasing opportunities for socially mediated learning through proximity, attention, and tolerance rather than demonstrable imitation. Overall, these behaviors broaden our understanding of sexual diversity within primate species and suggest that macaques, like other animals, exhibit a degree of sexual flexibility that extends beyond stereotypical mating rituals.

5.7 Social Bonds vs. Sexual Bonds

The distinction between social bonds and sexual bonds is pivotal in understanding the social behavior of the long-tailed macaque. While sexual bonds are typically transient and closely linked to reproductive cycles, social bonds are more enduring, shaped by factors such as kinship, cooperation, and the sharing of resources. Social bonds serve as the foundation for the macaque group's social structure, facilitating group cohesion, mutual support, and cooperative behaviors essential for survival (Cowlishaw & Dunbar, 1991; Dunbar, 2012). These bonds, often formed between related individuals, are vital for maintaining stability and order within the group.

In contrast, sexual bonds are more temporary, primarily influenced by the immediate reproductive needs of individuals rather than long-term social cooperation. These bonds can facilitate brief alliances between males and females, often driven by mating opportunities, but they do not extend beyond the reproductive context (Berman, 1982). The transient nature of sexual bonds underscores their role in fulfilling reproductive goals, in contrast to the more enduring social bonds that support group cohesion and stability.

The interplay between social bonds and sexual bonds illustrates the complex balance between individual reproductive strategies and the collective needs of the group. While sexual bonds serve to address the immediate reproductive interests of individuals, social bonds are crucial for the long-term survival and stability of the group. The dynamics of these bonds reflect the macaque's ability to navigate between individual and collective needs, ensuring both personal reproductive success and the group's overall success across generations.

In summary, Nicobar long-tailed macaques' social and reproductive behaviors are shaped by intricate interactions between dominance, cooperation, resource sharing, and sexual strategies. Males may dominate the hierarchical structure, but the females, through their social bonds and resourceful behaviors, play a pivotal role in shaping group dynamics. This interplay between social cohesion and sexual selection contributes to group stability and reproductive dynamics in Nicobar long-tailed macaques.

6 Tool Use and Culture, Discovering Newness and Local Vibes

Lasting Traits and Vanished Details

The investigation of tool use in non-human primates constitutes a critical domain for advancing our understanding of cognitive capacities, adaptive responses to ecological challenges, and the broader evolutionary pathways underpinning the emergence and transmission of culture. Over the past several decades, extensive empirical research has documented diverse manifestations of tool use across a wide range of primate taxa, spanning from great apes such as chimpanzees and orangutans to various monkey species exhibiting differing degrees of behavioral complexity (Carvalho & Almeida-Warren, 2019; Haslam et al., 2009, 2017; Pascual-Garrido et al., 2023).

Among these, long-tailed macaques (*Macaca fascicularis*) have emerged as a particularly compelling subject for study due to their notable behavioral flexibility and ecological versatility, which are exemplified in their sophisticated repertoire of tool-related behaviors. The present chapter provides a comprehensive examination of the distinctive tool-using behaviors exhibited by long-tailed macaques, with a focused analysis on the Nicobar long-tailed macaque, an insular subspecies inhabiting the unique ecological context of the Nicobar Islands.

Through a comparative lens, this chapter situates the tool-use strategies of the Nicobar long-tailed macaque within the broader framework of primate tool-use phenomena, highlighting both convergent and divergent patterns in cognitive engagement, cultural transmission, and environmental interactions. By doing so, it contributes to a nuanced understanding of how specific environmental pressures and social dynamics shape the innovation and perpetuation of tool use, thereby enriching current perspectives on primate cognition and the evolutionary underpinnings of culture.

6.1 Novel Behaviors

Long-tailed macaques exhibit a diverse and context-dependent repertoire of behaviors that can be characterized as 'novel,' emerging from their

DOI: 10.4324/9781003614579-6

dynamic interactions with the biotic and abiotic features of their habitats (Gumert & Malaivijitnond, 2013; Haslam et al., 2009; Tan et al., 2015; Watanabe et al., 2007). Such behaviors often arise as adaptive responses to specific ecological constraints and opportunities, as well as to the social structures within which these primates operate. This section examines the range of context-dependent behavioral responses observed among long-tailed macaques, with particular emphasis on those documented in the Nicobar Islands and situates them within the broader discourse on behavioral innovation and the evolution of cultural practices in non-human primates.

Field studies and experimental research have consistently documented innovative food-processing techniques among long-tailed macaques (Gumert et al., 2011; Mazumder & Kaburu, 2020; Pal et al., 2017; Tan et al., 2015). Notably, these primates have been observed using lithic materials, such as stones, as percussive tools to crack open nuts, shellfish, and other hard-shelled food items (Gumert & Malaivijitnond, 2013; Haslam et al., 2016; Tan et al., 2015). This behavior exemplifies their capacity for extractive foraging and highlights their ability to identify and exploit affordances in their environment to solve complex foraging challenges. Furthermore, the manipulation of objects to extract concealed or otherwise inaccessible food resources underscores their problem-solving skills, which are widely regarded as foundational to the emergence of cumulative cultural traditions in primate societies (Davidson, 2016).

Within the unique ecological context of Southeast Asia, the long-tailed macaque demonstrates a suite of behaviors not commonly reported among other subspecies. For instance, these macaques have been observed utilizing coral fragments as tools to probe substrates and access embedded invertebrate prey, a striking example of environmental opportunism shaped by the distinctive marine and coastal resources available on the islands (Luncz et al., 2019). Additionally, their use of leaves as absorptive implements to collect and transport water reveals a level of behavioral flexibility that reflects both environmental adaptation and cognitive engagement with the affordances of their insular habitat (Cenni et al., 2023). Notably, Cenni et al. (2023) documented similar behaviors in Balinese long-tailed macaques in Ubud, Indonesia, where over a four-year period researchers observed these primates engaging in tool-assisted water scooping using a variety of materials, including leaves, nuts, pits, and stones. Such behaviors demonstrate these macaques' remarkable behavioral plasticity and capacity for innovative problem-solving. Similar behaviors have been documented in other macaque species as well. For example, lion-tailed macaques (*Macaca silenus*) have been observed using leaves to collect drinking water from tree holes – a behavior comparable to that seen in apes (Fitch-Snyder & Carter, 1993). Recent research by Rosien et al. (2025) provides compelling evidence of advanced cognitive processing in long-tailed macaques. The study observed that these primates modulate their food-washing behavior in response to the diminishing returns of continued effort. Specifically, when the benefits of additional washing decreased, the macaques reduced their effort, indicating an understanding of cost-benefit trade-offs (Rosien et al., 2025). This behavior exemplifies a sophisticated level of decision-making, suggesting that long-tailed macaques can assess and respond to changing environmental contingencies. Such findings underscore the species' behavioral flexibility and cognitive adaptability, contributing to our understanding of the evolution of complex foraging strategies in primates.

In contrast to some conspecific populations in Southeast Asia that rely extensively on stone tools for extractive foraging, Nicobar long-tailed macaques exhibit a *distinct* and *context-dependent* repertoire of object- and tool-related behaviors shaped by their insular ecology and available materials. Pal et al. (2017) documented seven distinct complex manipulative behaviors within this subspecies, including six extractive foraging techniques and a unique dental hygiene practice. These behaviors encompass the use of both natural materials – such as leaves and sticks – and synthetic items, including plastic wrappers, reflecting an ability to opportunistically exploit both organic and anthropogenic resources. For instance, these macaques use leaves to wrap and wipe food items, which likely enhances cleanliness and may aid in removing toxins or debris. Sticks are employed to extract insects from crevices, demonstrating problem-solving skills and an understanding of cause-effect relationships. Particularly striking is their use of feathers and threads for dental flossing, indicating a level of self-care and bodily maintenance rarely observed among non-human primates. Pal et al. (2017) documented Nicobar long-tailed macaques using feathers and threads to floss their teeth, showcasing an opportunistic use of both natural and anthropogenic materials. Similarly, Mazumder and Kaburu (2020) reported the use of nylon threads for dental grooming, highlighting the macaques' adaptability in incorporating novel synthetic items into their hygiene repertoire. In a related context, Watanabe et al. (2007) described long-tailed macaques in Thailand employing human hair as dental floss, a behavior observed primarily in a human-provisioned setting and suggesting a degree of social learning, as adult females modified their flossing actions in the presence of infants. Together, these observations underscore the cognitive flexibility and cultural potential of macaques in developing and transmitting intricate self-care behaviors.

Such behaviors suggest not only sophisticated object manipulation but also the potential for cultural transmission of knowledge within social groups. Further enriching this perspective, Mazumder and Kaburu (2020) reported that Nicobar long-tailed macaques employ a wide variety of objects across multiple behavioral contexts, including foraging, hygiene,

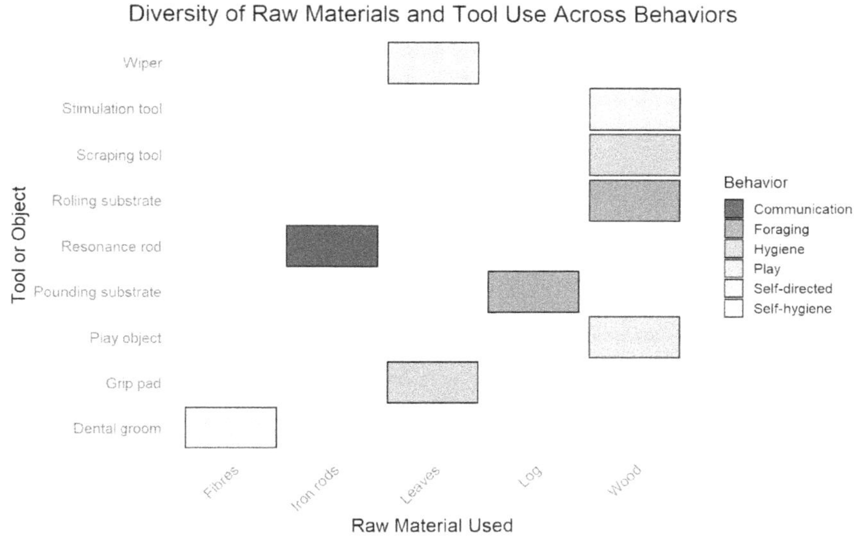

Figure 6.1 Summary of Tool or Object Types Used by Monkeys, Categorized by Raw Material and Associated Behavioral Context.

communication, play, and self-directed activities. Their study documented the use of eight distinct object types: resonance rods, play objects, rolling platforms, scraping tools, dental grooming implements, pounding substrates, leaves as grip pads and wipers, and stimulation tools (Figure 6.1). For example, sticks were observed being used as scraping tools to extract insects from bark or crevices, while leaves served as both grip pads for handling thorny or otherwise hazardous food items and as wipers for fur maintenance (Figure 6.2). Notably, Mazumder and Kaburu (2020) highlighted sex-based differences in tool-use frequencies, with males engaging in object manipulation more frequently than females, suggesting potential socio-ecological or cognitive factors influencing these patterns, an observation that aligns with findings from other populations (Gumert et al., 2011; Gumert & Malaivijitnond, 2013; Haslam et al., 2013; Tan et al., 2018).

Mazumder and Kaburu (2020) documented the use of sticks as stimulation tools by Nicobar long-tailed macaques in the Nicobar Islands. Their observations revealed that these macaques employ sticks not only for foraging but also for self-stimulation behaviors, indicating a multifaceted interaction with their environment. This behavior suggests a level of behavioral flexibility and adaptation to their ecological niche. In contrast,

Figure 6.2 A Male Nicobar Long-Tailed Macaque (*Macaca fascicularis umbrosus*) Uses a Scraping Tool, an Instance of Deliberate Object Manipulation Likely Aimed at Extracting Food or Processing a Surface.

Cenni et al. (2023) reported on the use of stone tools for self-stimulation by long-tailed macaques in Bali, Indonesia. The study highlighted that these macaques select specific stones and employ them for genital stimulation, a behavior that appears to be culturally transmitted within the population. Cenni's findings underscore the complexity of tool use in macaques, extending beyond utilitarian functions to include self-directed behaviors (Cenni et al., 2022).

Such behaviors not only enrich our understanding of macaque cognitive abilities but also emphasize the role of environmental context and cultural transmission in shaping primate tool-use repertoires. Comparing studies reveals both similarities and differences in tool use among macaque populations. These findings collectively highlight the ecological flexibility and cultural variation in macaque tool use, wherein materials – natural or anthropogenic – are repurposed to serve functions beyond subsistence, encompassing self-maintenance and potentially social communication.

However, the materials used, such as sticks in the Nicobar macaques and stones in the Balinese macaques, or feathers and nylon thread in Nicobar macaques, whereas human hair in Thailand reflects regional ecological variations and possibly differing cultural practices. Additionally, the social contexts and transmission mechanisms of these behaviors may differ, with potential influences from human interactions and provisioning in the Balinese population.

Comparatively, other macaque species exhibit tool-use behaviors shaped by their specific ecological contexts and the affordances of their respective environments. For instance, Burmese long-tailed macaques (*Macaca fascicularis aurea*) inhabiting coastal regions of Thailand are well-known for their use of stone tools to crack open sea almonds and oysters – a behavior demonstrating sophisticated extractive foraging and reflecting adaptation to their marine-influenced habitats (Falótico et al., 2017). In contrast, Japanese macaques (*Macaca fuscata*) have been observed engaging in stone-handling behaviors that appear primarily playful rather than strictly utilitarian, suggesting that not all object manipulation serves an immediate foraging function (Gunst et al., 2007). The tool and object-use repertoire of Nicobar long-tailed macaques encompasses a broad range of functional contexts, including extractive foraging, food processing, and self-maintenance. This diversity highlights behavioral flexibility without implying greater cognitive complexity relative to other macaque species that express different, ecologically structured tool-use strategies (Mazumder & Kaburu, 2020; Pal et al., 2017). This diversity in tool-use strategies among macaques underscores the profound influence of ecological pressures and resource availability on the emergence and maintenance of cultural behaviors in non-human primates. Moreover, it highlights the significance of insular environments as crucibles for behavioral innovation and the potential transmission of culturally learned practices within social groups (Figure 6.3).

Collectively, these findings underscore the complex and varied nature of tool use in Nicobar long-tailed macaques, distinguishing them from other macaque populations in Southeast Asia. The convergence of extractive foraging, hygiene-related, and social-object-use behaviors within this insular population offers valuable insights into the evolutionary underpinnings of primate cognition, cultural transmission, and ecological adaptability. Such observations highlight the importance of considering both environmental context and intraspecific variation when evaluating the origins and diversification of tool-use traditions among non-human primates. Here, Table 6.1 presents a synthesis of the diversity of tool- and object-use behaviors discussed throughout this chapter and provides a comparative reference for subsequent discussions of regional variation and ape–monkey contrasts.

Figure 6.3 An Adult Male Long-Tailed Macaque (*Macaca fascicularis umbrosus*) Engages in Extractive Foraging by Removing Bark from a Tree to Access Concealed Insects.

6.2 Macaque Culture

This section situates the Nicobar findings within broader debates on macaque cultural behavior, drawing primarily on comparative evidence from other macaque populations where social learning and the persistence of behavioral traditions have been more explicitly demonstrated. Building upon the diversity of behaviors outlined in the previous section, the concept of culture in macaques provides a useful comparative framework for examining how behavioral variants may emerge, stabilize, or disappear across populations under different ecological and social conditions. In non-human primates, culture is typically defined as socially learned behaviors that are transmitted within and between generations, forming stable traditions that persist over time (van Schaik et al., 2003; Whiten, 2020; Whiten et al., 1999). This concept challenges the traditional anthropocentric view that culture is exclusive to humans and highlights the evolutionary continuity of cultural processes across species. Among macaques, culture manifests most conspicuously in domains such as food processing, social interactions, and tool use, where behavioral variants are transmitted through social learning mechanisms rather than innate predispositions alone.

The concept of 'protoculture' – the transmission of behaviors without the mediation of complex symbolic communication – has gained widespread

Table 6.1 Comparative table of tool use in non-human primates

Species/Subspecies	*Tool Type*	*Material*	*Function/Use Context*	*Similar Behavior in Other Species*
M. f. umbrosus (Nicobar)	Stick	Wood	Insect extraction, scraping	Chimpanzees – termite fishing (*Pan troglodytes*) (Koops, 2020; Marchant, 1992)
M. fascicularis	Coral fragment	Coral	Probing substrates for marine invertebrates	Sea otters – rock use for shellfish cracking (Hall & Schaller, 1964; Haslam et al., 2019)
M. f. umbrosus	Leaf as grip pad or gloves	Leaf	Water collection, food wrapping, fur wiping	Chimpanzees – leaf sponging (*Pan troglodytes*) (Hockings et al., 2015)
M. f. umbrosus	Threads/nylon/ feathers	Organic/ synthetic	Dental hygiene – flossing	Japanese macaques – human hair for flossing (*M. fuscata*) (Leca et al., 2010)
M. f. umbrosus	Resonance rods	Iron rods found on buildings	Possible play by juveniles, or aggressive communication	Japanese macaques – stone handling as play (*M. fuscata*) (Huffman, 1996; Huffman et al., 2010)
M. f. umbrosus	Rolling platforms	Bark/leaves	Foraging behavior	Capuchins – object rolling and play (*Sapajus spp.*) (Falótico et al., 2018)
M. f. umbrosus	Pounding substrates	Bark/ground	Foraging behavior	Capuchins – stone pounding of nuts (Falótico et al., 2018)
M. f. umbrosus	Stimulation tools	Sticks	Genital or self-stimulation	*M. fascicularis* (Bali) – stone stimulation (Cenni et al., 2022)
M. f. aurea (Burmese)	Hammerstones & anvils	Stone	Cracking sea almonds, oysters	Chimpanzees – nut cracking (*Pan troglodytes*) (Benito-Calvo et al., 2015)

(*Continued*)

Table 6.1 (Continued)

Species/Subspecies	*Tool Type*	*Material*	*Function/Use Context*	*Similar Behavior in Other Species*
M. fuscata (Japanese macaque)	Stones	Stone	Play and social object manipulation	Seen in *M. fascicularis* (Bali) (unique play culture) (Cenni et al., 2024)
M. fuscata	Leaf as scoop	Leaf	Water scooping from tree holes	Similar to *M. f. umbrosus* and chimpanzee leaf sponging (Cenni et al., 2023; Hockings et al., 2015)
M. fascicularis (Thailand)	Floss	Human hair	Dental flossing	Similar to Nicobar macaques (using synthetic thread)
M. fascicularis (Bali)	Stone, leaf, shell	Natural materials	Water scooping/manipulation	Similar to long-tailed macaques' water tool use
Sapajus libidinosus (Bearded capuchin)	Stone + anvil	Stone	Nut cracking	Similar to *M. f. aurea* and chimpanzees
Pan troglodytes (Chimpanzee)	Stick, leaf sponge, spear	Plant material	Termite fishing, drinking, hunting	Multiple overlaps with macaques in leaf and stick use
Pongo spp. (Orangutan)	Leaf gloves, probing stick	Leaf/stick	Protection, insect probing	Similar leaf and stick use in *M. f. umbrosus*

acceptance in primatology (Price, 1973). Protoculture encompasses behaviors acquired through mechanisms such as imitation, emulation, and local enhancement, which enable individuals to acquire knowledge and skills by observing others (Galef, 1992; Perry & Manson, 2003; Whiten et al., 2005). This form of cultural transmission has been documented across various macaque species and is particularly evident in their tool-use traditions.

Long-tailed macaques are particularly instructive in this regard, as their broad geographic distribution and ecological diversity give rise to marked differences in behavioral repertoires among populations. For instance, populations inhabiting coastal regions of Thailand have developed sophisticated percussive tool-use behaviors to crack open sea almonds and oysters – a clear example of cultural adaptation to local ecological affordances (Falótico et al., 2017; Tan et al., 2015). In contrast, Japanese macaques engage in stone-handling behaviors that appear to be more play-oriented than strictly utilitarian, highlighting how not all object manipulations serve an immediate functional purpose but may also fulfill socio-cognitive roles (Huffman, 1996; Huffman & Quiatt, 1986).

The Nicobar long-tailed macaques provide a particularly compelling case study of primate culture within an insular environment. Their behavioral repertoire, encompassing extractive foraging, hygiene-related object use, and self-maintenance practices (Mazumder & Kaburu, 2020, 2021; Pal et al., 2017), exemplifies the interplay between ecological opportunities and the social transmission of novel behavioral expressions. Notably, Mazumder and Kaburu (2020) documented eight distinct object types employed by these macaques across multiple contexts, ranging from resonance rods and scraping tools to dental grooming implements and leaves used as grip pads – underscoring both the diversity of tool-use traditions and the potential for ecological adaptations within social groups. Furthermore, the documented sex-based differences in tool use among Nicobar long-tailed macaques, with males engaging more frequently in object manipulation, may reflect differential opportunities for social adaptation or divergent social roles within groups (Mazumder & Kaburu, 2020). Such findings align with broader observations in primate societies, where factors such as dominance rank, kinship, and social network centrality have been shown to influence opportunities for acquiring novel skills and transmitting them to others (Canteloup et al., 2020, 2021; Tan et al., 2018). The convergence of these socio-ecological factors likely contributes to the emergence and maintenance of behavioral traditions, reinforcing the notion that culture in non-human primates is both dynamic and context-dependent.

Comparative studies across macaque species further underscore the importance of social learning in the emergence of cultural behaviors. Japanese macaques, for instance, are well-known for their potato-washing and wheat-washing behaviors, initially innovated by a single individual and later adopted by other group members through social learning (Kawai, 1965). Similarly, stone-handling behaviors observed in Japanese macaques vary in form and function across populations, suggesting local traditions maintained by social learning (Gunst et al., 2007; Leca et al., 2012). These examples highlight the potential for cumulative cultural evolution in macaques, albeit at a more modest scale than in humans, with behavioral variants persisting and potentially elaborating across generations.

Social learning mechanisms underpinning these cultural behaviors are further evidenced in studies of social interactions and conflict management among macaques. For example, Aureli (1992) documented post-conflict behaviors in wild long-tailed macaques, revealing that victims of aggression exhibited increased self-directed scratching – a potential indicator of stress – and were more likely to be re-targeted by aggression. Significantly, reconciliatory behaviors, such as affiliative interactions between former opponents, were commonly observed, suggesting that macaques employ learned social strategies to repair relationships and mitigate the costs of aggression. These reconciliation behaviors are not isolated to a single population but have been consistently observed across both captive and wild settings, indicating that such practices are maintained through social learning and reinforced by group dynamics.

Building upon Aureli's seminal research on post-conflict reconciliation in macaques, which illuminated the mechanisms by which individuals repair social bonds to maintain group cohesion (Aureli, 1992), the recent study on despotic long-tailed macaques expands our understanding of prosocial behaviors within primate social systems (De Laat et al., 2024). Contrary to traditional views that despotic hierarchies limit cooperative interactions, this new research demonstrates that individuals engage in group-beneficial behaviors even in the absence of direct personal gain, indicating a capacity for prosociality that transcends dominance structures. These findings suggest that social learning plays a crucial role not only in the transmission of material culture but also in fostering complex social behaviors that enhance group stability. Together, these studies underscore the multifaceted nature of macaque culture, encompassing both behavioral strategies for conflict resolution and cooperative interactions, thereby enriching our comprehension of the evolutionary roots of sociality in non-human primates.

Then again, in addition to social behaviors, the material culture of long-tailed macaques has been a subject of increasing interest. A study by Malaivijitnond et al. (2007) documented the use of axe-shaped stones by wild long-tailed macaques in Thailand to crack open rock oysters and other shellfish (Malaivijitnond et al., 2007). This behavior not only demonstrates the macaques' ability to use tools but also suggests a level of cultural transmission, as such practices are observed consistently within certain groups and are likely learned through social observation and imitation.

The emergence of material culture among non-human primates, particularly long-tailed macaques, has inspired a growing body of archaeological research aimed at identifying, excavating, and interpreting primate-modified sites (Falótico et al., 2017; Haslam et al., 2016, 2017). One of the most significant insights into macaque material culture has emerged from studies of stone-tool-assisted foraging among Burmese long-tailed macaques inhabiting the coastal regions of Thailand (Falótico et al., 2017; Haslam et al., 2016). These primates have been documented using stone hammers and anvils to process hard-shelled invertebrates and sea almonds, behaviors that leave behind distinctive archaeological signatures such as pitted and fractured stones, shell middens, and spatially organized 'worksites' (Falótico et al., 2017). Systematic excavation at these sites has revealed repeated use of specific tool-use locations over time, providing robust evidence for the persistence and potential transmission of tool-assisted foraging behaviors.

For example, Gumert et al. (2009) identified repeated use of intertidal zones as tool-use 'hotspots,' characterized by clusters of hammerstones and anvils, often associated with discarded shells and other food remnants. These sites offer a rare opportunity to examine the durability and dispersal of material culture in a non-human primate population. Moreover, the spatial distribution and stratigraphic layering of these sites demonstrate that macaque tool use is not ephemeral but leaves a tangible, archaeologically detectable record that can persist across multiple generations of macaques (Haslam et al., 2009).

While the Nicobar long-tailed macaques have yet to be systematically investigated through formal excavation, preliminary observations of their tool-use behaviors suggest the potential for discovering comparable archaeological sites. For instance, Pal et al. (2017) and Mazumder and Kaburu (2020) documented a variety of tool-use behaviors among this subspecies, including the use of leaves, sticks, and even anthropogenic materials for food processing, hygiene, and social functions. However, the apparent lack of stone tool use in this population could potentially complicate excavation efforts, as organic tools – such as wood, bone, or plant fibers – are highly perishable and decompose far more readily than lithic artifacts, leaving a biased archaeological record (Cnuts & Rots, 2024; Hiscock, 1985). Nevertheless, systematic excavations of select sites could still reveal important evidence of past tool-use behaviors that have ceased to exist due to environmental changes or ecological pressures. Such excavations might uncover ephemeral yet informative traces of these behaviors – such as modified plant materials, discarded tool fragments, or spatial concentrations of use-wear debris – similar to those identified at sites in Thailand (Haslam et al., 2016). The significance of such excavated sites lies in their potential to: (1) establish behavioral continuity, linking contemporary observations of tool use with long-term patterns of repeated site use and tool deposition – thus reinforcing the argument for cultural transmission and persistence (Haslam, 2012); (2) provide comparative frameworks, enabling direct comparisons between macaque populations (e.g., Burmese vs. Nicobar macaques) and with other primates (e.g., chimpanzees, capuchins), thereby refining our understanding of material culture in non-human primates (Falótico et al., 2017; Haslam et al., 2016; Mercader, 2002); and (3) illuminate evolutionary processes by offering insights into the cognitive, social, and ecological factors that shape the emergence of material culture – potentially informing hypotheses about early hominin tool use (Call & Tennie, 2009; Tennie et al., 2016; Whiten et al., 1999). Collectively, these findings would provide invaluable insights into the behavioral repertoire and cultural transmission among insular macaque populations.

Taken together, these findings underscore that macaque culture is multifaceted and context-dependent, shaped by the interplay between ecological factors, resource availability, and social structures. The Nicobar long-tailed macaques, through their unique insular adaptations, highlight the evolutionary significance of cultural flexibility and demonstrate how behaviors can become embedded within group traditions. These observations emphasize the importance of considering both ecological and social dynamics in the study of primate culture, and they support the value of comparative approaches that examine cultural evolution across diverse macaque populations.

The case of the Nicobar long-tailed macaque illustrates how ecological challenges can generate population-specific behavioral variants that are compatible with socially mediated processes observed in other macaque populations, although direct evidence for social learning in this subspecies remains limited and warrants further investigation. Their use of both natural materials (e.g., leaves, sticks) and anthropogenic materials (e.g., plastic wrappers, nylon threads) reflects adaptability and opportunistic exploitation of locally available resources. The repeated occurrence of these behaviors across groups is consistent with patterns reported in other macaque populations. Importantly, these behaviors are not interpreted as indicators of heightened cognitive complexity relative to other macaques, but rather as context-dependent expressions shaped by material availability, ecological constraints, and repeated individual interactions with environmental affordances.

This body of research not only enriches our understanding of cultural evolution in macaques but also offers valuable insights into the evolutionary origins of culture in humans. Future research that integrates behavioral observations with genetic, ecological, and demographic data will help to elucidate the cognitive and social mechanisms driving cultural behaviors in macaques, ultimately bridging the gap between primate and human cultural evolution.

6.3 Tool Use vs. Object Use

6.3.1 What Distinguishes These Long-Tailed Macaques from Other Macaques in the Nicobar Island Chain?

The Nicobar archipelago harbors diverse populations of long-tailed macaques, yet significant distinctions emerge in their patterns of interaction with objects and tools across different islands. Among these, the population inhabiting Great Nicobar stands out for its documented engagement in tool-assisted extractive foraging, a behavior that is comparatively less evident or altogether absent among conspecifics on neighboring islands, i.e., Katchall and Little Nicobar (Mazumder, 2022; Mazumder & Kaburu, 2020).

These behavioral differences appear to be closely linked to local ecological contexts, including the availability of specific food resources, substrate composition, and patterns of anthropogenic influence. On Great Nicobar, for example, the presence of hard-shelled nuts, screw pines, certain insect prey, and diverse vegetation types likely provides both the impetus and the opportunity for the development and maintenance of extractive tool use (Mazumder & Kaburu, 2021). Such a relationship between resource characteristics and tool-use innovation aligns with broader findings in primatology, where the emergence of tool-assisted foraging has been linked to the affordances and challenges of particular environments (Biro et al., 2013; Sanz & Morgan, 2007)

In contrast, macaque populations on adjacent islands often display object manipulation behaviors – such as play, exploration, or social display – without transitioning these actions into functional tool use. This distinction underscores the importance of differentiating between object use, which may encompass a range of manipulative behaviors, and true tool use, which involves the deliberate modification or selection of an object to achieve a specific, goal-directed outcome (Mazumder & Chauhan, 2025; Shumaker et al., 2011). The latter reflects a higher level of flexibility, including sensitivity to object affordance and repeated manipulation of material properties, that sets the Great Nicobar population apart within the Nicobar chain.

Moreover, the consistent use of tools by Great Nicobar macaques invites comparative analysis with other primate tool users, such as chimpanzees (*Pan troglodytes*) and capuchins (*Sapajus* spp.), highlighting the potential convergent evolution of tool use in response to similar ecological and social pressures (Babitz, 1999; Barrett et al., 2018; Falótico & Ottoni, 2015; Sanz & Morgan, 2007). Investigating the specific factors driving this divergence – whether cognitive, ecological, or socio-cultural – remains an important avenue for understanding the evolution of complex behaviors in primates.

Taken together, these insights underscore that while object use is widespread among macaques, the emergence of systematic tool use in Great Nicobar's population reveals a unique intersection of environmental affordances and adaptive capacities that warrants further study. Formal excavation and long-term observation could elucidate the processes through which tool traditions emerge and persist in these insular primates, contributing significantly to our understanding of cultural evolution in nonhuman primates.

6.4 Exploring Novelty and Regional Dynamics

Patterns of tool use and object interaction among long-tailed macaques across the Nicobar archipelago illustrate the dynamic interplay between ecological pressures, social adaptation, and demographic processes. Regional variation within these insular populations underscores the importance of local environmental contexts in shaping behavioral repertoires, as well as the potential for behavioral innovations to emerge or disappear over time.

The geographic isolation characteristic of island systems fosters unique socio-ecological conditions that can catalyze the emergence of novel behaviors, including tool use and object manipulation. For example, differences in habitat structure – such as canopy continuity, substrate types, and resource distributions – directly influence foraging strategies and the potential for tool-assisted behaviors. Such ecological heterogeneity can foster divergent behavioral solutions among neighboring populations, even within the same species (Gumert et al., 2013; Haslam et al., 2016; Pal et al., 2017; Tan et al., 2015; Watanabe et al., 2007).

Moreover, the ephemeral nature of certain behavioral traits highlights the delicate balance between innovation and cultural persistence. Behaviors that initially arise in response to specific environmental challenges may fail to stabilize within a population if social transmission mechanisms – such as observational learning or tolerance for close-range foraging – are insufficiently robust (McGrew et al., 2001). Conversely, strong social networks and repeated exposure to particular behaviors can facilitate the consolidation of innovations into enduring cultural traditions, reinforcing group-level behavioral distinctiveness (Luncz et al., 2017, 2019).

Comparative analyses of Nicobar macaque populations thus provide an invaluable framework for exploring the processes by which ecological variables, demographic factors, and social learning intersect to shape the emergence and maintenance of behavioral diversity. Recognizing these regional dynamics is critical not only for understanding the adaptive significance of tool use but also for elucidating the broader evolutionary mechanisms underpinning primate cultural evolution. Future research integrating behavioral, ecological, and genetic data will be essential to unravel these complex interactions and to refine our models of cultural transmission in non-human primates.

6.5 Ape Tool Use vs. Monkey Tool Use

Tool-use behaviors in non-human primates have long been recognized as windows into the cognitive and cultural capacities that underlie complex social learning and problem-solving. A critical distinction emerges when comparing apes and monkeys in terms of the diversity and sophistication of their tool-use repertoires, as well as the mechanisms underlying their acquisition and maintenance. The ape-monkey distinction used here is heuristic rather than taxonomic and is intended to contrast broad patterns of tool-use organization, not to imply evolutionary equivalence, hierarchy, or representativeness across diverse primate lineages. The species discussed are not treated as proxies for broader taxonomic categories, but as illustrative case studies demonstrating how distinct evolutionary histories and ecological pressures shape material engagement in primates.

Among apes, particularly chimpanzees (*Pan troglodytes*) and orangutans (*Pongo* spp.), tool-use behaviors often exhibit a high degree of technical complexity and contextual flexibility (Arroyo et al., 2016; Bandini et al., 2021; Boose et al., 2013; Luncz et al., 2018; Shumaker et al., 2011). These species have been documented employing tool sets – multiple tools used sequentially to achieve a single goal – and demonstrating a capacity for tool modification to adapt to specific functional demands (Boesch & Boesch, 1989). Moreover, their tool-use traditions are frequently embedded within rich social learning frameworks, including vertical (parent-offspring), oblique (adult to juvenile), and horizontal (peer) transmission pathways, contributing to the emergence of stable cultural traditions that persist across generations (Aplin, 2016; Barrett et al., 2017; Boesch & Boesch, 1989; Lonsdorf, 2005).

In contrast, monkey tool use – while diverse – typically exhibits a more ecologically opportunistic character. Long-tailed macaques, for example, predominantly use tools in extractive foraging contexts, often capitalizing on environmental affordances such as accessible substrates or food resources that can be processed using simple implements (Luncz et al.,

2017; Mazumder & Kaburu, 2021). In monkey species, the reliance on social learning for the acquisition of tool-use behaviors appears comparatively attenuated relative to that observed in apes. Empirical evidence indicates that juveniles reared by humans often acquire tool-related skills through interactions with human caregivers, whereas those reared in naturalistic conditions predominantly learn from high-ranking conspecifics within their social groups (Fredman & Whiten, 2008). This pattern suggests that observational learning and social tolerance, while present, play a more limited role in the propagation and maintenance of tool-use traditions in monkeys compared to the more robust social transmission characteristic of ape populations.

Nevertheless, recent studies challenge the traditional dichotomy that positions ape tool-use as categorically more advanced than that of monkeys. Evidence from insular populations of long-tailed macaques, including the Nicobar long-tailed macaque, reveals instances of tool-use behaviors that demonstrate a surprising degree of functional sophistication and potential cultural adaptations. Such cases suggest that under certain ecological and demographic conditions, monkeys may develop tool-use traditions that approximate the complexity and stability observed in apes. Table 6.2 provides a comparative overview of tool-use characteristics across apes and monkeys, highlighting broad cognitive, social, and ecological dimensions reported in the literature. The table is intended as a synthetic comparison rather than as evidence for specific learning mechanisms within any single population.

These findings emphasize the importance of conceptualizing primate tool use not as a discrete or uniformly expressed trait, but as a continuum emerging from the interaction of ecological opportunity, material availability, social tolerance, developmental exposure, and individual experience. Within this framework, tool use is best understood as a flexible behavioral response shaped by local conditions rather than as a fixed marker of cognitive capacity. Such an approach allows variation, incompleteness, and even apparent absence of tool use to be interpreted as informative outcomes of ecological and social processes, rather than as deficits.

Advancing this perspective will require integrative research strategies that combine longitudinal behavioral observations with detailed ecological mapping, demographic monitoring, and, where possible, genetic and archaeological data. Together, these approaches can help clarify how material engagement is initiated, how it is constrained or facilitated by social and environmental contexts, and why certain behaviors persist across generations while others remain ephemeral or disappear entirely. Importantly, this integrative lens also enables the investigation of behavioral loss and attenuation, processes that are rarely documented but are

Table 6.2 Comparative characteristics of tool use in apes and monkeys: cognitive, social, and ecological dimensions

Aspect	*Ape Tool Use*	*Monkey Tool Use*
Representative Species	Chimpanzees (*Pan troglodytes*), Orangutans (*Pongo* spp.)	Long-tailed macaques (*Macaca fascicularis*), Capuchins (*Sapajus* spp.)
Tool Complexity	High complexity; use of multiple tool sets and modification of tools for specialized tasks (Boesch et al., 2009)	Generally simpler; single tools often unmodified, primarily used for extractive foraging and basic functions (Falótico & Ottoni, 2015; Pal et al., 2017)
Cognitive Processes	Advanced causal understanding; planning and foresight in tool manufacture and use (Baber, 2003; Mosley & Haslam, 2016)	Opportunistic use with functional understanding but less evidence of foresight or complex planning (Tan et al., 2015)
Social Learning and Transmission	Strong reliance on social learning mechanisms, including vertical, horizontal, and oblique transmission pathways (Lonsdorf, 2005)	More limited social learning; tool use may be individually learned, context-dependent or socially facilitated but with less robust transmission (Mazumder & Kaburu, 2020; Tan et al., 2018)
Cultural Traditions	Well-documented, stable, and regionally distinct tool-use cultures (Haslam et al., 2018)	Emerging evidence of cultural traditions, though often less stable and more variable (Falótico & Ottoni, 2015; Mazumder & Kaburu, 2020; Tan et al., 2018)
Functional Contexts	Diverse, encompassing foraging, social displays, and environmental manipulation (Shumaker et al., 2011)	Primarily extractive foraging, hygiene, and occasionally social interactions (Mazumder & Kaburu, 2020; Shumaker et al., 2011)
Ecological Drivers	Tool use often driven by complex environmental challenges requiring innovation (Luncz et al., 2018)	Tool use largely driven by immediate ecological opportunities and resource availability (Luncz et al., 2017)

(*Continued*)

Table 6.2 (Continued)

Aspect	*Ape Tool Use*	*Monkey Tool Use*
Representative Examples	Termite fishing with modified sticks, nut cracking with stone hammers, leaf sponging (Boesch & Boesch, 1989; Koops, 2020; Lonsdorf, 2005; Mercader, 2002)	Stone use for shellfish cracking, stick use for probing, Use of wooden platform as rolling or pounding substrate, use of human hair, nylon thread and bird feathers for hygiene (Malaivijitnond et al., 2007; Mazumder & Kaburu, 2020; Pal et al., 2017; Watanabe et al., 2007)
Evolutionary Significance	Reflects advanced cognitive and social capacities, providing models for early hominin technological evolution	Demonstrates broader taxonomic distribution of tool use, emphasizing the role of ecology and sociality in behavioral innovation

likely central to understanding the evolutionary dynamics of material traditions in primates.

Within this broader comparative and process-oriented framework, Nicobar long-tailed macaques assume a particularly informative role in primate research. Their significance does not lie in the breadth or complexity of their tool-use repertoire, but in their capacity to illuminate how innovation, regional specificity, ecological constraint, and behavioral attrition interact over time. As an insular population inhabiting rapidly changing socio-ecological landscapes, they offer a rare opportunity to examine how material traditions are shaped, maintained, modified, or lost under conditions of environmental variability and anthropogenic pressure.

By foregrounding these dynamics, the Nicobar macaques contribute to a growing body of work that moves beyond ape-centric and progress-oriented narratives of tool evolution, and toward a more nuanced understanding of material behavior as a historically contingent and environmentally embedded process. In doing so, they provide a valuable empirical context for rethinking the evolutionary pathways through which primate traditions arise, stabilize, or disappear, and for refining theoretical models that link behavior, ecology, and material engagement across primate taxa.

7 Conservation and the Future of Tool-Use Behavior and Cultural Heritage in Nicobar Macaques

7.1 The Great Nicobar Project

The relevance of the Great Nicobar Project to this monograph lies not only in its scale, but in its spatial overlap with ecologically specific behaviors documented in earlier chapters – particularly coastal foraging zones, intertidal tool-use sites, and forest-edge habitats repeatedly used by Nicobar long-tailed macaques. Unlike generalized habitat loss, such landscape transformations directly threaten the physical contexts in which tool use is expressed, learned, and maintained. This section, therefore, examines the project not simply as a development initiative, but as a driver of behavioral erosion. Accordingly, this chapter treats tool-use behavior in Nicobar long-tailed macaques not as an isolated behavioral trait, but as an ecologically embedded process that depends on specific substrates, foraging zones, and socially tolerated spaces. By examining how large-scale infrastructure development disrupts these conditions, the chapter demonstrates how conservation planning that overlooks behavioral landscapes risks preserving populations while eroding the ecological contexts necessary for behavioral persistence.

7.1.1 The Project

The Great Nicobar Island Development Project (GNIDP), also referred to as the *Great Nicobar Development Plan*, marks a landmark infrastructural intervention on the southeastern tip of Great Nicobar Island. Conceptualized by the National Institution for Transforming India Aayog (NITI Aayog) and implemented by the Andaman & Nicobar Islands Integrated Development Corporation (ANIIDCO), the project seeks to reposition Great Nicobar as a critical node in India's Indo-Pacific maritime strategy (Govt. of India, 2022a; Sekhsaria, 2024a). It aims to bolster regional connectivity, economic activity, and national security through an integrated framework involving urban, maritime, aviation, and energy infrastructure.

DOI: 10.4324/9781003614579-7

Sanctioned on 11 November 2022 by the Ministry of Environment, Forest and Climate Change (MoEFCC) under the title *Holistic Development of Great Nicobar Island*, the project is estimated to require an investment of ₹72,000 crore (approximately USD 8.7 billion), with a projected implementation horizon spanning over three decades (Sekhsaria, 2024a). One of the project's flagship components is the proposed International Container Transshipment Terminal (ICTT) at Galathea Bay, with a standalone allocation of ₹40,000 crore (~USD 4.85 billion), intended to rival major regional ports such as Colombo and Singapore. The project's total footprint will involve the diversion of approximately 166.10 square kilometers of land, about 18% of the island's total area of 910 km^2, including over 130 km^2 of primary evergreen rainforest, one of the last remaining intact tropical forest systems in India (Sekhsaria, 2024a).

The proposed phase-wise land allocation reveals a carefully staged progression. Phase I encompasses 72.20 km^2, followed by 54.34 km^2 in Phase II, and 39.56 km^2 in Phase III. A detailed breakdown shows that the ecotourism sector accounts for the largest dedicated land use (40.55 km^2), followed by green and recreational areas (23.87 km^2), and coastal regulatory zones (20.02 km^2) (Sekhsaria, 2024b) (For complete breakdown see Table 7.1). Core infrastructure includes:

- The Galathea Bay Transshipment Port (6.92 km^2).
- A dual-use greenfield airport (8.45 km^2).
- A power plant complex (0.39 km^2).
- And expansive residential (15.08 km^2) and institutional zones (5.77 km^2).

Township and area development, encompassing housing, utilities, and commercial facilities, represents the most land-intensive component – covering approximately 149.60 km^2. This development is expected to permanently transform the island's southeast quadrant, historically characterized by low-density indigenous settlements and forested landscapes.

Source: Minutes of the 30th meeting of the Expert Appraisal Committee, 22–23 August 2022 (Sekhsaria, 2024b).

7.1.2 *Chronology of Development and Dissent*

The making of the Great Nicobar Project (Table 7.2): the developmental trajectory of the Great Nicobar Island (GNI) over the last two decades illustrates a deeply contested intersection of ecological conservation, national security, and large-scale infrastructure planning. The contemporary saga begins as early as 2005, with post-tsunami scientific assessments such as those by the Earthquake Engineering Research Institute (EERI)

Table 7.1 Phase-wise land allocation for the Great Nicobar Development Project

Land Use	*Phase I (sq km)*	*Phase II (sq km)*	*Phase III (sq km)*	*Total (sq km)*
Airport	8.45	0	0	8.45
ICTT	4.5	2.42	0	6.92
Power plant	0.39	0	0	0.39
Residential	6.78	6.45	1.85	15.08
Commercial – Mixed Use	1.59	1.08	0.85	3.52
Commercial – Office	0.18	0	0	0.18
Commercial – Tourism & Hospitality	2.53	2.48	0.01	5.01
Commercial – Coastal Tourism	1.31	2.16	4.72	8.19
Institutional	1.26	4.52	0	5.77
Industrial and Logistic	0.13	4.68	0.07	4.88
Eco-tourism	16.51	10.62	13.42	40.55
Green and Recreational	9.56	10.92	3.39	23.87
Utilities – Others	0.13	0.21	0.08	0.42
Utilities – Water Reservoir	1.27	1.3	0	2.57
Jetty	0.1	0	0	0.1
Defence Land	12.61	0	0	12.61
Special Use (Defence Purpose)	0.43	0	0	0.43
Special Use (National Memorial)	0.08	0	0	0.08
Special Use (Pemayya Buffer Area)	0	0.69	3.83	4.52
Coastal Regulatory Zone (CRZ)	4.39	6.82	8.81	20.02
Total	**72.2**	**54.34**	**39.56**	**166.1**

documenting significant subsidence of land in the Nicobar region. The fragile ecological recovery of the island's coastline was confirmed by subsequent surveys, notably a 2011 Wildlife Institute of India (WII) study that recorded the return of leatherback turtles to Galathea Bay, a critical nesting habitat (Jadeja et al., 2016; Sekhsaria, 2024a; Shanker et al., 2011).

According to published records, the first formal signal of infrastructural ambition came in 2015, when the Port Management Board called for an Expression of Interest to develop a Transshipment Port (TSP) and Free Trade Warehousing Zone in the Nicobar Islands. This was followed by increasing momentum under the central government's Sagarmala Programme, with AECOM's 2016 technical report cautioning that a cruise facility was preferable to a deep-sea port due to the region's ecological sensitivity. Nevertheless, policy traction intensified. By 2019, the Island Coastal Regulation Zone (ICRZ) Notification categorized GNI as a high-priority development zone, and NITI Aayog's publication *Transforming the Islands* presented a SWOT (Strengths, Weaknesses, Opportunities, and Threats) analysis promoting 'holistic development,' focusing heavily on eco-tourism, port infrastructure, and a dual-use airport.

Table 7.2 Overview of the Project's decision-making processes (Sekhsaria, 2024a, 2024b)

Date	*Event*	*Category*
1956–2004	The Andaman and Nicobar Protection of Aboriginal Tribes Regulation (ANPATR) is promulgated and subsequently amended through 2004.	Legislation
Apr. 2005	The Earthquake Engineering Research Institute (EERI) documents land subsidence in Great Nicobar Island following the 2004 Indian Ocean earthquake.	Scientific Report
Feb. 2007	The Wildlife Institute of India (WII) undertakes a post-tsunami survey assessing the impact on flora and fauna, focusing on the Nicobar megapode.	Scientific Report
Oct. 2011	**Post-tsunami** – WII's conservation genetics study records the reappearance of leatherback turtles at Galathea Bay.	Scientific Report
2012	**MoEF Report (COP11)** – The Ministry of Environment and Forests (MoEF) includes a section on the Great Nicobar Biosphere Reserve in India's Biosphere Reserves report.	Policy Report
22 May 2015	**Date Unspecified** – A policy addressing the Shompen Tribe of Great Nicobar Island is released.	Policy Document
Jul. 2016	**AECOM's Technical Report** – AECOM concludes that only a cruise facility is feasible at Great Nicobar Island; port construction is deemed unsuitable.	Technical Report
Nov. 2016	The Ministry of Shipping and the Indian Ports Association finalize the Sagarmala report, incorporating AECOM's July 2016 analysis.	Technical Report
Jan. 2017	**Date Unspecified** – Research highlights Galathea Bay as a critical nesting site for leatherback turtles in Nicobar.	Scientific Report
08 Mar. 2019	**ICRZ Notification** – GNI is classified as a Group I oceanic island exceeding 1,000 sq. km.	Legal Notification
22 May 2019	**FAC Recommendation** – The Forest Advisory Committee recommends Compensatory Afforestation (CA) for forest diversion projects in high-forest-cover states/UTs to be implemented in forest-deficient states/UTs.	Policy Decision
26 Dec. 2019	**WAPCOS Tender** – WAPCOS issues a tender for a traffic study for the transshipment port at Galathea Bay.	Tender/Contract

(*Continued*)

Table 7.2 (Continued)

Date	*Event*	*Category*
2019	**NITI Aayog Initiative** – NITI Aayog commissions a video report capturing stakeholder perspectives on GNI's development, later posted on the Andaman & Nicobar Collective's YouTube channel.	Technical report
22 Jun. 2020	**Project Interest** – Three firms, including Adani, respond to the A&N Port Management Board's Expression of Interest (EoI) for the transshipment port project; the Government of India grants in-principle approval.	Tender/ Contract
11 Aug. 2020	**Prime Minister's Announcement** – The Prime Minister announces the transshipment port project for GNI, with an investment of ₹10,000 crore.	Government Announcement
04 Sep. 2020	**Tribal Welfare Committee** – The Director of Tribal Welfare forms an empowered committee to examine NITI Aayog's proposals for land acquisition and development in GNI.	Policy Decision
Sep. 2020	**NITI Aayog** issues a statement addressing bidders' queries	Technical report
07 Sep. 2020	**NITI Aayog RfP** – NITI Aayog issues a Request for Proposal (RfP) for preparing a Master Plan for Holistic Development of GNI.	Tender/Contract
28 Oct. 2020	**ESZ Draft Notification** – A draft notification proposes a 0–1 km Eco-Sensitive Zone (ESZ) around Campbell Bay National Park.	Legal Notification
Nov. 2020	**Adani Ports Presentation** – The India Brand Equity Foundation's presentation by Adani Ports references the proposed transshipment port under 'Recent Trends and Studies.'	Tender/Contract
04 Nov. 2020	**NCZMA Approval** – The National Coastal Zone Management Authority (NCZMA) approves recategorization of GNI from Group I to Group II at its 41st meeting.	Regulatory Decision
01 Jan. 2021	**Notification** – Official recategorization of GNI from Group I to Group II is published.	Legal Notification
19 Jan. 2021	**National Marine Turtle Action Plan** – MoEFCC releases the National Marine Turtle Action Plan.	Policy Document
25 Jan. 2021	**Denotification** – Galathea Wildlife Sanctuary and Megapode Wildlife Sanctuary are denotified.	Legal Notification

(*Continued*)

Table 7.2 (Continued)

Date	*Event*	*Category*
02 Mar. 2021	**Nodal Agency Designation** – ANIIDCO is designated as the nodal agency for the holistic development of GNI.	Administrative Decision
06 May. 2021	**RTI Response** – NITI Aayog replies to an RTI stating no vision document was prepared for GNI's development plans.	RTI Response
14 Jun. 2021	**MoEFCC ToR** – MoEFCC issues Terms of Reference (ToR) for Environmental Impact Assessment (EIA) to ANIIDCO.	Environmental Clearance
28 Jul. 2021	**ZSI Report** – The Zoological Survey of India (ZSI) prepares a report on conservation of flora and fauna as part of the EIA.	Scientific Report
18 Nov. 2021	**IPZ Amendment** – Integrated Coastal Regulation Zone (ICRZ) 2011 is amended to permit gas-based power plants in ICRZ III areas.	Legal Notification
24 Dec. 2021	**Land Acquisition** – Local officials (Tehsildar, Patwari, AC) sign an undertaking for land acquisition for airport and road projects.	Administrative Action
27 Jan. 2022	**Public Hearing** – A public hearing takes place in Campbell Bay.	Stakeholder Engagement
04 Mar. 2022	**Supreme Court Directive** – The Supreme Court advises a petitioner challenging Galathea WLS denotification to approach the High Court.	Legal Order
07 Mar. 2022	**ANIIDCO Submissions** – ANIIDCO submits its final EIA report and application for environmental and CRZ clearance.	Environmental Clearance
29 Apr. 2022	**UN CERD Inquiry** – The UN Committee on the Elimination of Racial Discrimination (CERD) contacts GoI regarding vulnerable tribal groups in A&N.	International Inquiry
08 Jun. 2022	**Nodal Port Agency** – The Ministry of Shipping declares Syama Prasad Mookerjee Port Trust as the nodal agency for the proposed port.	Administrative Decision
27 Oct. 2022	**Forest Clearance** – MoEFCC grants Forest Clearance for the GNI Holistic Development project.	Environmental Clearance
11 Nov. 2022	**Environmental Clearance** – MoEFCC grants Environmental Clearance for the same project.	Environmental Clearance
18 Nov. 2022	**MoTA Response** – The Ministry of Tribal Affairs (PVTG Division) states they lack information on A&N's tribal communities and redirects the RTI query to the Ministry of Home Affairs.	Governance/RTI

(*Continued*)

Table 7.2 (Continued)

Date	*Event*	*Category*
22 Nov. 2022	**Tribal Council Withdrawal** – The Tribal Council withdraws its NoC for forest diversion, citing lack of consent and information concealment.	Governance/ Forest Rights
22 Nov. 2022	**ZSI Publication** – ZSI releases a book on the faunal biodiversity of GNI.	Scientific Report
24 Nov. 2022	**NGT Challenge** – Debi Goenka and Conservation Action Trust challenge the Forest Clearance at the National Green Tribunal (NGT).	Legal/ Environmental
27 Nov. 2022	**Concerns Raised** – Former GoI Secretary EAS Sarma highlights project approval without prior consultation with the National Commission for Scheduled Tribes (NCST).	Governance/ Policy
08 Dec. 2022	**Parliamentary Question** – Jawhar Sircar questions forest land diversion in GNI in the Rajya Sabha.	Legislative Query
21 Dec. 2022	**ANIIDCO RfP** – ANIIDCO issues an RfP for township master plan consultants, referencing a 'vision document.'	Planning/Tender
11 Jan. 2023	**NGT Hearing** – The first NGT Kolkata hearing challenges both Forest and Environmental Clearances.	Legal/ Environment
28 Jan. 2023	**Port Call for EoI** – Syama Prasad Mookerjee Port invites EoIs for Phase I of the ICTT project at Galathea Bay.	Port Development
08 Feb. 2023	**CFA Deliberations** – The Centre for Financial Accountability hosts online deliberations on the GNI project (Day 1).	Public Engagement
Feb. 2023	**CFA Report** – CFA publishes a report on the GNI project.	NGO Report
09 Feb. 2023	**CFA Deliberations** – Day 2 of CFA's online discussions takes place.	Public Engagement
14 Feb. 2023	**ANIIDCO Affidavit** – ANIIDCO files a counter-affidavit in the NGT.	Legal
17 Feb. 2023	**MHA Declaration** – The Ministry of Home Affairs designates 10 sensitive installations in A&N under the Official Secrets Act.	National Security
22 Feb. 2023	**MoEFCC Affidavit** – MoEFCC files a counter-affidavit in the NGT.	Legal
07 Mar. 2023	**CFA Session** – CFA hosts another session on the GNI project.	NGO Report
09 Mar. 2023	**NCST Consultation** – NCST conducts a stakeholder consultation on GNI.	Governance/ Tribal Rights
24 Nov. 2023	**DPR Status** – The Detailed Project Report (DPR) for the transshipment port nears finalization, with bids anticipated next year.	Port Development

(Continued)

Table 7.2 (Continued)

Date	*Event*	*Category*
15 Nov. 2023	**Ministerial Visit** – Minister Sarbananda Sonowal (MoPSW) visits the Galathea port site to review progress.	Government Review
22 Nov. 2023	**SACON Advertisements** – SACON advertises for positions in macaque, megapode, crab, endemic birds, and hazard assessments related to the airport project.	Conservation/ Research
23 Nov. 2023	**Afforestation Rejection** – The Municipal Corporation of Gurgaon denies land allocation for Compensatory Afforestation.	Governance
09 Jan. 2024	**Shipping Ministry Statement** – The Shipping Ministry announces that construction at Galathea Bay Port will begin within three months; in-principle approval from the Department of Expenditure is confirmed.	Port Development
07 Feb. 2024	**International Advocacy** – Genocide scholars from 13 countries write to the President of India urging project cancellation due to potential impacts on the Shompen community.	Indigenous Rights
20 Feb. 2024	**Presidential Visit** – The President of India visits GNI, meets the tribal community, and emphasizes balancing development and the environment.	Government Visit
24 Feb. 2024	**NCST Comment** – The NCST Acting Vice-Chairperson calls international criticism of the GNI project a conspiracy.	Political Commentary
04 Mar. 2024	**MoEFCC Report Withheld** – MoEFCC denies the High Powered Committee report's disclosure under the RTI Act.	Legal/ Environment
14 Mar. 2024	**SIA Notification** – A notification is issued for a Social Impact Assessment (SIA) for 404.8194 hectares in Gandhinagar and Shastrinagar villages.	Land Acquisition
15 Apr. 2024	**Lok Sabha Elections** – Lok Sabha elections occur in A&N; contrary to reports, the Shompen had already voted in 2014.	Election/Media
19 Apr. 2024	**SIA Award** – ANIIDCO awards ₹2,173,475 to Probe Research & Social Development Pvt Ltd for the SIA for the airport.	Tender/Contract
23 Apr. 2024	**Ex-Servicemen Commemoration** – Settlers in GNI commemorate 55 years since the first Punjabi settlers' arrival in 1969.	Commemoration
30 Apr. 2024	**Lieutenant Governor Review** – The Lieutenant Governor reviews the GNI Township draft master plan.	Governance/ Planning

(*Continued*)

Table 7.2 (Continued)

Date	*Event*	*Category*
02 May 2024	**Draft Plan Review** – Top officials meet the Shompen PVTG to review the draft master plan.	Governance/ Tribal Rights
14 May 2024	**NGT Direction** – The NGT directs MoEFCC to respond to construction activity in prohibited CRZ-I areas.	Legal/ Environment
Feb. 2025	**SACON Baseline Studies** – SACON initiates a two-year baseline ecological study to monitor project impacts.	Research and Monitoring
Apr. 2025	**Survival International Report** – Survival International submits a report to the UN CERD urging the project's cancellation due to risks to the Shompen tribe.	International Advocacy
Apr. 2025	**ICTP Notification** – The Galathea Bay site is notified as India's 13th major port.	Infrastructure Development
Apr. 2025	**Environmental Management Budget** – SACON's conservation plan budget of ₹230.8 crore over 30 years is integrated into the project's management plan (details not publicly disclosed).	Conservation Funding
Apr. 2025	**Monitoring Committees** – Monitoring committees begin overseeing the project's compliance with environmental and tribal welfare stipulations.	Project Oversight
May 2025	**Shompen Extinction Concerns** – The project is criticized for potentially leading to Shompen cultural and ecological extinction.	Environmental and Cultural Impact
Nov. 2025	National Green Tribunal reviews High-Powered Committee report; selected sections withheld citing national security	Legal / Environmental Oversight
Late 2025 (ongoing)	Project framed as a strategic Indo-Pacific and national security infrastructure initiative centered on Galathea Bay	Government / Strategic Policy
Late 2025	Projected cost revised to approximately ₹72,000–92,000 crore; final implementation phases advanced under ANIIDCO	Infrastructure Development
Late 2025	Scientific studies continue to document new and poorly known taxa within Great Nicobar Island	Scientific Report
Jan. 2026	Reports indicate pressure on Great Nicobar Tribal Council regarding land surrender for project implementation	Governance / Tribal Rights
Continues	Continued concern regarding deforestation scale (~8.5 lakh trees), coral reef damage, and tectonic vulnerability	Environmental Risk Assessment

Despite these visions, a parallel stream of scientific and legal red flags began to surface. A 2016 leatherback turtle distribution study reaffirmed Galathea Bay as the most important nesting site in the Indian Ocean. Simultaneously, internal communications revealed shifts in land-use categorization and environmental zoning (Jadeja et al., 2016). In 2020, the National Coastal Zone Management Authority (NCZMA) recategorized GNI from Group I (>1000 km²) to Group II, thereby diluting protective environmental thresholds. Later that year, the Wildlife Sanctuaries (WLS) in Galathea and Megapode habitats were de-notified, clearing the way for land acquisition.

By 2021, the project entered its accelerated implementation phase. The ANIIDCO was appointed as the nodal agency, and multiple undertakings were signed – including admissions that ANIIDCO lacked an environmental policy and that tribal welfare departments would provide any required legal exemptions. Public hearings were announced and held in Campbell Bay, despite critiques that local communities were not meaningfully consulted and environmental impact assessments were insufficient.

In the lead-up to official project sanction, symbolic ecological gestures emerged. On 6 October 2022, Haryana Chief Minister Manohar Lal Khattar announced the establishment of a 10,000-acre Aravalli Jungle Safari, just three weeks before the Great Nicobar project was granted Forest Clearance by the Ministry of Environment, Forest and Climate Change (MoEFCC) (Sekhsaria, 2024a). Notably, Khattar's announcement suggested that funding would be sourced from ANIIDCO, revealing a strategic alignment between compensatory afforestation policy and mainland political optics. This move was later revealed to have occurred while internal documents acknowledged the clearance of nearly a million trees on Great Nicobar.

The final clearances were followed by a revocation of the No Objection Certificate (NoC) by the Tribal Council, which argued that the initial consent was obtained without transparency and under misleading pretenses. Further compounding the controversy, the Ministry of Tribal Affairs (MoTA) declared in an RTI (Right to Information) response that it had no information on the tribal communities of the Andaman & Nicobar Islands, effectively displacing accountability to the Ministry of Home Affairs and NITI Aayog.

Legal challenges swiftly followed. In November 2022, the Conservation Action Trust filed an appeal in the National Green Tribunal (NGT) contesting the Forest Clearance. Simultaneously, international scrutiny grew, with the UN Committee on the Elimination of Racial Discrimination (CERD) seeking clarification from the Government of India regarding the protection of Particularly Vulnerable Tribal Groups (PVTGs) and compliance with the Forest Rights Act (FRA).

Figure 7.1 A Macaque Stands Atop a Freshly Chopped Tree, Its Posture Capturing a Moment of Stillness Amid Signs of Habitat Disturbance.

7.1.3 Forest Conservation within the Development Agenda

The Great Nicobar Island Development Project received environmental clearance in 2022, permitting the felling of approximately 850,000 trees (Figure 7.1). Subsequent ecological assessments indicate that the total number of mature trees ultimately affected may exceed two million (Sekhsaria, 2024a). This scale of deforestation represents not merely a loss of forest cover, but a profound disruption to canopy microclimates, soil stability, hydrological cycles, and carbon sequestration systems. The anticipated ecological degradation will accelerate carbon emissions through the destruction of both biomass and soil organic carbon pools, intensifying the island's vulnerability to climate variability.

Of particular concern is Galathea Bay, the proposed site for the container transshipment terminal. This ecologically sensitive area lies within the Great Nicobar Biosphere Reserve, recognized by UNESCO in 2013 for its global conservation value. Galathea National Park, which anchors this zone, hosts one of the last significant nesting beaches of the endangered leatherback sea turtle (*Dermochelys coriacea*) in the Indian Ocean (Sekhsaria, 2024a; Sharma, 2024; Swaminathan, 2021). The region's ecological integrity, highly susceptible to port-induced stressors such as sedimentation, noise, and artificial light, is poised to be irreversibly compromised by infrastructure expansion.

In parallel, the project's expansive footprint encroaches upon lands traditionally inhabited and stewarded by Nicobarese and Shompen communities. These indigenous populations maintain deep-rooted cultural, spiritual, and subsistence relationships with the forest and coastal ecosystems targeted for conversion. Their territories encompass hunting grounds, sacred groves, and shifting cultivation plots, now facing fragmentation or displacement. This raises acute concerns regarding livelihood security, cultural erosion, and the marginalization of indigenous rights within India's broader development agenda.

Although statutory mechanisms such as Environmental Impact Assessments (EIAs), biodiversity offsets, and compensatory afforestation are formally in place, their efficacy in moderating large-scale environmental transformation is increasingly questioned. One of the most contested aspects of this project is the proposal to offset the loss of over 130 km^2 of primary rainforest through tree planting in mainland regions like Haryana's Aravalli Hills. This compensatory approach, though legally sanctioned, suffers from an ecological and spatial disconnect: the humid tropical evergreen forests of Great Nicobar are biologically and functionally incompatible with the semi-arid, degraded, dry deciduous landscapes of northern India. Consequently, such offsets offer no meaningful replacement for the island's biodiversity and ecosystem services. Instead, they reduce complex habitats to simplified carbon metrics, commodifying forest cover for developmental legitimacy.

This ecological and political contradiction came into sharp focus in October 2022, when Haryana Chief Minister Manohar Lal Khattar announced plans to develop India's largest jungle safari, spanning 10,000 acres (approximately 40.5 $kilometers^2$) across the Gurugram and Nuh districts. Touted as a flagship initiative in conservation-based tourism, the announcement was made on 6 October 2022, just weeks before the Great Nicobar project received forest clearance on 11 November 2022. Notably, this proposal surfaced after internal communications had already circulated within government channels regarding the impending clearance of nearly one million trees in the Andaman and Nicobar Islands. The timing of the safari project, framed as an act of ecological stewardship, suggests a strategic alignment with the central government's compensatory afforestation policies, even as it highlights the vast geographical, ecological, and socio-cultural disconnect between donor and recipient sites of environmental restoration.

This case exemplifies a structural flaw in India's conservation governance, the spatial and political decoupling of ecological accountability. By relocating restoration efforts to regions disconnected from the sites of environmental loss, the policy framework privileges developmental symbolism over ecological substance. It externalizes environmental costs, allowing high-value ecosystems like the Great Nicobar to be sacrificed under the pretext of offsetting elsewhere. Such practices risk converting conservation into a ritualized bureaucratic exercise, where restoration is quantified but ecological integrity is not restored. Ultimately, the Great Nicobar Island Development Project crystallizes the dilemma facing contemporary environmental governance in India. Situated at the crossroads of geostrategic ambition and ecological fragility, it offers a cautionary tale of how large-scale infrastructural aspirations can imperil one of the subcontinent's last great wildernesses and the indigenous communities who inhabit it.

While development impacts are typically evaluated in terms of biodiversity loss or human displacement, they rarely account for the disruption of non-human behavioral traditions tied to specific landscapes. The following sections examine how infrastructure development on Great Nicobar threatens not only species survival, but also the ecological conditions required for the persistence of tool-use behaviors documented in Nicobar macaques.

7.2 Overlooked Risks and Governance Challenges

Despite its ambitious vision for infrastructural development and regional connectivity, the Great Nicobar Project is marred by significant loopholes and procedural shortcomings that compromise its stated goals of sustainable development and environmental stewardship. These oversights are not merely administrative but have profound implications for biodiversity conservation, indigenous rights, and the long-term ecological resilience of the Nicobar Islands.

7.2.1 Inadequate Environmental Impact Assessment (EIA)

One of the most critical loopholes lies in the fragmented and narrowly scoped Environmental Impact Assessments conducted for the project. Instead of a comprehensive, cumulative impact analysis that integrates terrestrial, coastal, and marine ecosystems, the EIAs have adopted a piecemeal approach that fails to account for the interconnectedness of these habitats. This oversight is particularly concerning given the region's status as part of the globally significant Andaman–Nicobar Biosphere Reserve and its designation as a UNESCO Biosphere Reserve (Sekhsaria, 2024b). The lack of rigorous baseline data on species distribution, habitat connectivity, and ecosystem services undermines the credibility of the proposed mitigation measures.

7.2.2 Neglect of Indigenous Communities and Free, Prior, and Informed Consent (FPIC)

The project's planning framework exhibits a profound deficit in the meaningful inclusion of indigenous communities – primarily the Shompen and Nicobarese – whose customary lands and cultural landscapes intersect directly with the proposed development zones. Although the Forest Rights Act (2006) and international conventions such as the United Nations Declaration on the Rights of Indigenous Peoples (UNDRIP) mandate Free, Prior, and Informed Consent (FPIC), the project approvals proceeded without adequate community consultation or culturally appropriate participatory mechanisms. This procedural gap threatens to erode indigenous sovereignty and exacerbates socio-economic vulnerabilities already intensified by previous displacement events and limited healthcare infrastructure.

7.2.3 Absence of Systematic Archaeological Surveys

Despite the well-established role of the Andaman and Nicobar archipelago as a corridor of ancient human migrations and cultural interactions across Southeast Asia and the Indian subcontinent, comprehensive archaeological surveys of Great Nicobar Island remain conspicuously absent. The sparse and fragmentary data that exist are largely confined to preliminary reconnaissance studies, with limited scope for understanding the region's deep-time human occupation, trade networks, and cultural landscape dynamics (Agrawal, 1969; Arora, 2010; Cooley, 2014; Macnaughtan, 2013; Mishra et al., 2024). This lacuna in the archaeological record severely undermines the ability of planners and conservationists to develop contextually informed and culturally sensitive management strategies that integrate both tangible and intangible heritage values.

7.2.4 Overlooked Ethnoarchaeological Significance of Non-Human Tool-Use Behavior

Equally troubling is the neglect of non-human primate tool-use sites, specifically those associated with Nicobar macaques, which represent a unique dimension of the island's cultural and evolutionary heritage. Field studies have indicated the presence of complex tool-use behaviors in these macaque populations, including the use of rudimentary tools for foraging and extractive tasks. However, these behaviors, their spatial distribution, and their relationship to the broader forest ecology have not been systematically mapped or integrated into archaeological heritage management frameworks. This oversight marginalizes a critical component of the island's biodiversity and cultural heritage nexus, ignoring the evolutionary and cultural significance of primate tool-use traditions that have coexisted with indigenous human communities for millennia.

The conservation relevance of these behaviors lies not in their rarity alone, but in their spatial specificity. Tool-use sites reflect stable foraging locations, material affordances, and social tolerance zones that are not interchangeable across landscapes. Importantly, the loss of such sites would not merely reduce behavioral diversity, but would erase the material and spatial record necessary for primate archaeological investigation, foreclosing future opportunities to examine how tool-use traditions emerge, stabilize, or disappear in non-human primates.

The absence of robust archaeological documentation – both human and non-human – creates a governance vacuum that allows large-scale infrastructure projects to proceed without adequate consideration of potential impacts on irreplaceable heritage sites. This oversight not only jeopardizes the preservation of critical archaeological resources but also risks eroding the cultural transmission processes that sustain indigenous knowledge systems and ecological stewardship practices. Moreover, it challenges India's commitments under national and international frameworks, including the UNESCO Convention on the Protection of the Underwater Cultural Heritage (2001) and the United Nations Convention on Biological Diversity (1993), both of which emphasize the protection of cultural and natural heritage in development planning.

7.2.5 Overlooked Seismic and Geodynamic Risks

Geologically, Great Nicobar Island is situated within an active accretionary prism of the Andaman–Sumatra subduction zone, an area highly susceptible to seismic hazards, including earthquakes and tsunamis. Recent studies (Verma et al., 2023) have documented the presence of complex deformation structures such as thrust faults, pop-up structures, and schuppen zones (schuppen structure typically develops in compressional tectonic settings, such as mountain belts, where rocks are deformed by intense horizontal stress. It's commonly seen in thin-skinned tectonics where rocks are stacked up in slices or sheets) in the tuff beds of Shastri Nagarm, Great Nicobar Island.

Janaki Andharia and her team at the Tata Institute of Social Sciences have repeatedly raised alarms, highlighting in their comments on the draft EIA report that the island has recorded 444 earthquakes over the last decade (Arvind, 2024). They argue that the container terminal project should be reassessed considering these risks. Additionally, with sea levels continuing to rise, many island nations may be forced to relocate or reevaluate their prior economic risk assessments. Given this precarious situation, investing roughly ₹40,000 crore (approximately USD 4.8 billion) to build port infrastructure along this coastline appears both risky and imprudent.

Moreover, the broader Andaman–Nicobar region lies in close proximity to the active volcanic arc of the Andaman–Sumatra system, which includes Barren Island, the only confirmed active volcano in South Asia, and Narcondam, currently classified as dormant but geologically young (Bennis, 2023; Global Volcanism Program 2025). In addition to these magmatic centers, the islands also host several mud volcanoes, most prominently at Baratang and across Middle and North Andaman, which periodically erupt mud, water, and gases in response to tectonic activity (Press Trust of India, 2023). The continued seismicity of this arc underscores the volatile geological setting in which Great Nicobar is located. Although the volcanoes themselves are not on Great Nicobar, their activity signals ongoing tectonic and magmatic processes in the surrounding region, adding another dimension of uncertainty to large-scale infrastructure development. Moreover, paleotsunami deposits along the Andaman coast (e.g., Badabalu), analyzed by Malik et al. (Malik et al., 2019), record multiple high-magnitude tsunami events over the past ~8,000 years, underscoring the region's vulnerability to not only seismic but also tsunami hazards.

Looking ahead, climate change poses compounding threats to island chains such as the Nicobars. Rising sea levels and increasing storm surges are projected to cause coastal erosion, saltwater intrusion, and eventual submergence of low-lying areas (Figure 7.2) (Carruthers et al., 2024). These processes endanger not only human settlements but also archaeological sites that preserve the cultural history of the Nicobarese and Shompen communities, as well as critical habitats for endemic and endangered species (Anderson et al., 2017). Among these are the Nicobar long-tailed macaques, whose foraging zones, tool-use sites, and nesting areas are often located along intertidal and coastal zones that are highly vulnerable to inundation (Mazumder & Kaburu, 2021; Narasimmarajan & Raghunathan, 2012; Pal et al., 2018; Velankar et al., 2016, 2024). The dual pressures of climate change and infrastructure expansion thus threaten to erase both cultural heritage and biological diversity in one of India's most fragile ecological frontiers.

Yet, the project's risk evaluations fail to adequately consider these geological complexities in the design of crucial facilities, including ports, airports, and coastal settlements. The lack of thorough geotechnical studies, slope stability analyses, and tsunami hazard assessments raises significant concerns about the resilience of these infrastructures to natural disasters.

7.2.6 Compensatory Afforestation and Biodiversity Offsets

The project's mitigation strategies largely rely on compensatory afforestation and biodiversity offsetting, with proposed tree planting schemes located outside the island (e.g., in Haryana's Aravalli region). Such measures are ecologically unsound, as they fail to replicate the unique biotic assemblages, soil conditions, and microclimatic regimes of the Great Nicobar forests. The ongoing destruction of habitats on the Nicobar

Figure 7.2 A Stark Reminder of Coastal Vulnerability: This Eroded Shoreline Bears the Scars of a Past Tsunami, Where Powerful Waves Stripped Away Soil and Vegetation, Exposing Tree Roots and Destabilizing the Coastal Slope.

Islands is putting endemic species such as the Nicobar tree shrew (*Tupaia nicobarica*) and the Nicobar megapode (*Megapodius nicobariensis*) at serious risk. Moreover, the irreplaceable nature of these site-specific habitats – including critical nesting beaches for leatherback turtles (*Dermochelys coriacea*) and the culturally significant tool-use sites of the Nicobar long-tailed macaque (*Macaca fascicularis umbrosus*), renders offsetting an inadequate strategy for genuine conservation (Sekhsaria, 2024a).

7.2.7 *Neglected Energy Vulnerabilities and Sustainability Challenges*

Energy is a vital input for the daily functioning, economic growth, and overall well-being of any society, and this is especially true for the Andaman and Nicobar Islands. Energy consumption influences multiple development sectors including livelihoods, agriculture, health, education, and gender equity. Currently, approximately 90% of the islands' electricity generation depends on diesel fuel, which is costly ranging between ₹12 and ₹14 per kilowatt-hour. This heavy reliance on diesel translates to an enormous daily consumption of about 250 kiloliters, costing roughly ₹450 crores annually (approximately 60 million USD). Such expenditure not only strains the islands' economy but also contributes significantly to environmental degradation (Sonamuthu, 2019).

Despite the islands possessing substantial potential for renewable energy sources such as wind, solar, biomass, and small hydroelectric power, the ongoing development projects have largely failed to prioritize or adequately invest in these cleaner alternatives. The Kalpong Hydroelectric Project stands out as a notable example of renewable energy adoption, offering electricity at significantly lower costs than diesel generation and signaling a shift toward sustainability in the northern and middle regions of the archipelago. However, this progress remains limited, and much of the islands continue to grapple with aging infrastructure, frequent disruptions in fuel supply, and inefficiencies that hamper reliable energy delivery (Sonamuthu, 2019).

The current consumption patterns further reveal an imbalance compared to the mainland: domestic electricity use in the islands accounts for about 47.33% of total consumption (versus 22% nationally), while commercial use is around 26.32% (compared to 8% nationally) (Sonamuthu, 2019). These figures suggest high energy demand in non-industrial sectors, which adds pressure to an already fragile system. Without urgent and strategic investment in renewable energy infrastructure to reduce diesel dependence, the islands risk perpetuating an unsustainable energy model that undermines economic resilience and environmental health.

7.2.8 Unaddressed Freshwater Scarcity and Infrastructure Deficiencies

Water scarcity is a chronic and pressing issue in the Andaman and Nicobar Islands, particularly during the dry season from January to May, when many islanders face severe shortages. The project currently underway is estimated to require a staggering 86,600 kiloliters of water per day (KLD) to support its operations, of which approximately 45,000 KLD is planned to be sourced as freshwater from surface reservoirs that have yet to be constructed (Ramakrishna, 2022). This dependence on newly planned but unbuilt infrastructure raises serious concerns about the project's feasibility and sustainability.

Compounding this, the project's plans do not presently include clear provisions for the disposal, treatment, or safe recycling of waste and remnant water generated from its processes. The lack of an integrated water management strategy that addresses wastewater treatment and reuse not only jeopardizes freshwater availability but also poses risks of pollution and ecological damage.

Currently, the islands rely predominantly on a single major water source – the Dhanikhari Dam – which creates a dangerous dependency given the region's location in seismic Zone 5, a highly active earthquake zone (Sanjib, 2023). Any disruption to this dam due to seismic events could severely impact the water supply for large populations. Islanders have long demanded the creation of a dedicated *Jal* Board department responsible for comprehensive water management throughout the year. Such a body would ideally oversee the construction of additional water storage facilities, including large ponds and reservoirs in remote and vulnerable areas, to serve as emergency water sources during dry months.

The failure to diversify water sources, implement rainwater harvesting, develop groundwater recharge infrastructure, and incorporate sustainable wastewater management into project planning threatens to perpetuate annual water crises that disrupt communities, agriculture, and local ecosystems. Without decisive administrative action and infrastructural investment, the freshwater challenges facing the Andaman and Nicobar Islands will intensify, undermining the viability of large-scale development projects and the well-being of island residents.

7.2.9 *Institutional Fragmentation and Governance Deficits*

The governance architecture overseeing the Great Nicobar Project suffers from institutional fragmentation, with overlapping mandates among the Ministry of Environment, Forest and Climate Change, the NITI Aayog, the Andaman and Nicobar Administration, and various private stakeholders. This fragmented governance structure inhibits the effective integration of ecological and cultural dimensions into planning, monitoring, and enforcement. Moreover, the lack of an independent oversight mechanism or statutory conservation trust limits accountability and transparency, thereby increasing the risk of unmitigated ecological degradation.

Collectively, these loopholes highlight the urgent need for a more integrated, participatory, and scientifically robust approach to development in Great Nicobar. Without addressing these shortcomings, the project threatens to undermine not only the ecological integrity of the island but also the cultural heritage and rights of its indigenous inhabitants. A re-evaluation of the project through the lens of ecological justice, seismic risk preparedness, and indigenous rights is imperative for fostering a truly sustainable development paradigm.

7.3 Toward an Integrated Mitigation Strategy: A Feasible Path Forward

Considering the aforementioned procedural and substantive shortcomings, it is imperative to propose a mitigation strategy that is both scientifically rigorous and operationally feasible within the existing socio-political context. The following integrated mitigation plan aims to reconcile the urgent need for biodiversity conservation, cultural heritage preservation, and sustainable development on Great Nicobar Island. This strategy emphasizes participatory planning, interdisciplinary collaboration, and adaptive management, ensuring that conservation imperatives are not merely symbolic but embedded in the project's design, implementation, and governance.

7.3.1 *Comprehensive Cumulative Impact Assessment (CCIA)*

A key priority is the immediate commissioning of a Comprehensive Cumulative Impact Assessment (CCIA) that holistically integrates terrestrial, coastal, and marine ecosystems, cultural heritage (encompassing both human and non-human elements), water resource management, energy sustainability, and seismic risk factors (Declerck et al., 2022; Keskinen & Kummu, 2010; Piet et al., 2021). This CCIA must consolidate baseline data on species distribution, habitat connectivity, indigenous cultural landscapes, and archaeological resources, with specific consideration of Nicobar macaque tool-use localities as behavioral indicators of habitat continuity, resource stability, and social tolerance thresholds. The CCIA should be carried out by an independent, interdisciplinary consortium comprising ecologists, archaeologists, seismologists, anthropologists, and representatives from indigenous communities.

7.3.2 *Establishment of a Participatory Conservation Governance Framework*

A legally mandated Participatory Conservation Governance Framework (Jain, 2015; Sullivan, 2019) should be instituted to ensure that decision-making authority is shared equitably among relevant stakeholders, including indigenous Nicobarese and Shompen communities. This framework would operationalize Free, Prior, and Informed Consent (FPIC) protocols in accordance with the Forest Rights Act (2006) and international conventions (e.g., UNDRIP). The governance structure should include an independent Conservation Oversight Committee, comprising scientists, indigenous representatives, civil society organizations, and legal experts, to monitor compliance with conservation objectives and mitigation measures.

7.3.3 *Systematic Archaeological and Ethnoarchaeological Survey Prior to Development*

Effective mitigation in the context of large-scale development on the Nicobar Islands must begin with a recognition that landscapes are not only ecological spaces but also repositories of accumulated human and non-human histories. Archaeological and ethnoarchaeological approaches are therefore essential for identifying forms of heritage that are spatially persistent yet often overlooked by conventional environmental assessments. In addition to prehistoric and historic human occupation, these approaches are uniquely suited to documenting non-human primate behavioral landscapes, particularly where repeated activities leave material traces over time.

Prior to the initiation of any construction or land modification, a systematic archaeological and ethnoarchaeological survey should be undertaken across the project area. Such a survey must extend beyond the documentation of human settlement remains to include the identification and mapping of locations associated with non-human primate tool use and object-mediated foraging. Repeated use of specific sites by macaques may result in modified substrates, reused platforms, or accumulations of objects that constitute material signatures of behavior. Recording these features using archaeological methodologies allows planners to identify areas of long-term behavioral significance that may not be evident through short-term biological surveys alone.

Ethnoarchaeological components of the survey should further incorporate engagement with local and indigenous communities, whose place-based knowledge can provide critical insights into long-standing patterns of landscape use, ecological change, and human–animal interactions. Such perspectives are particularly valuable for identifying culturally significant zones that overlap with macaque habitats, reinforcing the understanding that human and non-human histories on the islands are deeply entangled rather than separable. Integrating these datasets allows mitigation planning to address heritage in a holistic sense, rather than privileging one form of heritage at the expense of another.

Complementing these living knowledge systems, archaeological research in the Nicobar Islands has previously demonstrated the value of midden deposits for reconstructing long-term patterns of landscape use, resource exploitation, and site fidelity in human populations (Cooper, 1993; Prakash, 2024). Such deposits are widely recognized as material indicators of repeated activity at particular locations and are routinely incorporated into heritage assessments and management plans. While non-human primate tool-use localities do not constitute middens in a formal archaeological sense, they similarly represent places where material traces accumulate through repeated use over time. Recognizing this methodological parallel does not equate human and non-human behavior, but rather underscores the utility of archaeological approaches for identifying and managing behaviorally significant places across species.

The results of these surveys should be formally embedded within Heritage Management Plans that guide project design and implementation. These plans must move beyond descriptive inventories to serve as operational documents that classify sites by cultural and behavioral significance, outline clear avoidance or buffering strategies, and establish binding protocols for site protection.

Where surveys identify areas of particular importance, such as repeatedly used tool-use localities or sites of overlapping human and non-human heritage, these zones should be designated as no-development areas or, where appropriate, as Community Conserved Areas (CCAs). Such designations enable development to proceed in less sensitive zones while safeguarding places that are critical for long-term behavioral continuity.

Importantly, archaeological and ethnoarchaeological mitigation should not be treated as a one-time exercise confined to the preconstruction phase. All data generated through surveys should be archived and revisited throughout the lifespan of the project, providing baseline information against which future changes can be assessed. In this way, archaeological documentation becomes an ongoing component of adaptive management, ensuring that heritage protection remains responsive to emerging impacts rather than static and symbolic.

7.3.4 *Primatological and Behavioral Mitigation Measures*

While archaeological approaches allow for the identification of behaviorally significant places, primatological mitigation focuses on the biological and ecological processes that enable such behaviors to emerge and persist. In the case of Nicobar long-tailed macaques, tool use does not appear to be a generalized species-wide trait, but rather a spatially constrained and ecologically contingent behavior tied to particular substrates, foraging zones, and environmental contexts. Mitigation strategies that prioritize population size or habitat extent alone are therefore insufficient, as they risk preserving animals while simultaneously undermining the conditions necessary for the maintenance of locally expressed behavioral traditions.

Primatological mitigation must begin with the identification of behaviorally critical microhabitats. These include recurrent foraging areas, stable platforms or substrates used for object manipulation, and spatial configurations that facilitate repeated access to particular resources. Although such features may occupy relatively small areas, their functional importance can be disproportionately large. Disturbance or removal of these microhabitats may disrupt behavioral expression even when broader habitat remains intact.

Equally important is the protection of linkages between behaviorally significant sites. Macaque tool use and foraging practices are embedded within daily movement patterns that connect sleeping sites, feeding areas, and object-use locations. Fragmentation of these pathways through roads, fencing, or increased human activity can reduce opportunities for repeated site use and, over time, erode behavioral persistence. Mitigation planning must therefore maintain landscape connectivity and avoid alterations that restrict access to key areas, even if such alterations appear minor from a purely infrastructural perspective.

Development-related disturbances should also be regulated with explicit reference to macaque behavior. Generic wildlife guidelines often fail to account for the temporal and spatial specificity of primate activities. Noise, artificial lighting, and sustained human presence near behaviorally significant sites may alter activity patterns or discourage site reuse. Establishing disturbance thresholds informed by behavioral observation, rather than by species presence alone, allows mitigation measures to be calibrated to actual risk rather than assumed tolerance.

Monitoring programs associated with the project should extend beyond demographic indicators to include systematic documentation of behavior. This may involve recording the frequency and context of object manipulation, assessing continued reuse of known sites, and tracking changes in foraging strategies over time. Success should not be measured solely by the continued presence of macaques in the landscape, but by the persistence of behavioral patterns that are locally characteristic and ecologically meaningful.

Finally, primatological mitigation must remain adaptive. Behavioral change may occur gradually and may only become apparent over extended periods. Management frameworks should therefore allow for the modification of project operations if monitoring indicates behavioral disruption, including the expansion of protected zones or the adjustment of activity schedules. By embedding behavioral data into post-development review processes, mitigation becomes a dynamic tool capable of responding to unanticipated impacts rather than a static set of prescriptions.

Taken together, archaeological and primatological mitigation measures offer complementary pathways for safeguarding both places and processes. Archaeology provides the tools to identify and protect spatially persistent traces of behavior, while primatology clarifies the ecological and social conditions under which such behaviors are expressed and maintained. Their integration ensures that conservation and development planning does not reduce macaques to population units alone, but recognizes them as agents embedded within historically layered landscapes. In doing so, mitigation efforts can move beyond the preservation of species survival toward the protection of locally sustained behavioral traditions that constitute an often overlooked dimension of biodiversity.

7.3.5 Integrated Infrastructure Planning for Resilience

Energy, Water, and Seismic Risk Considerations: Given the documented geodynamic risks in the Andaman and Nicobar Islands (Sekhsaria, 2024a; Verma et al., 2023), it is imperative that all infrastructure projects incorporate comprehensive seismic risk mitigation measures. These should include slope stability analyses, tsunami evacuation planning, and earthquake-resistant construction techniques to ensure resilience against natural hazards. To oversee this critical integration, an independent geotechnical task force must be established to review and guide project designs. Alongside seismic risk management, addressing the islands' pressing energy and water challenges is essential for sustainable development. Transitioning from diesel-based power generation to renewable energy sources such as solar, wind, small hydro, and biomass can significantly reduce environmental impact and operational costs. Enhancing energy efficiency and modernizing electrical infrastructure, combined with microgrid and energy storage solutions, will further improve energy reliability. Water security requires year-round management through a dedicated water authority, expanded water storage via ponds and reservoirs, rainwater harvesting, and wastewater treatment with reuse practices. All water infrastructure should be designed with seismic resilience in mind, supported by emergency plans to secure supply during crises. Engaging local communities in conservation efforts will strengthen the sustainability of these integrated strategies. By holistically incorporating seismic risk assessments alongside robust energy and water management into infrastructure planning, the project can better safeguard the islands' environment and communities against both natural and human-made threats.

7.3.6 In-Situ Conservation Over Off-Site Afforestation

Rather than relying on compensatory afforestation in ecologically dissimilar regions, the project should prioritize in-situ habitat restoration on Great Nicobar itself. This includes rehabilitating degraded habitats, protecting existing forest patches, and ensuring connectivity between forest fragments critical to the survival of endemic species and tool-using macaque populations. Special attention should be given to protecting nesting beaches for leatherback turtles and culturally significant tool-use habitats of Nicobar long-tailed macaques.

7.3.7 Interdisciplinary Monitoring and Adaptive Management

An Interdisciplinary Monitoring and Adaptive Management program should be established to regularly assess the ecological, cultural, and socioeconomic impacts of project activities. This program should include biodiversity surveys, cultural heritage audits, and community-based monitoring, with transparent reporting to the public and independent oversight bodies. Adaptive management frameworks should allow for iterative adjustments to project plans based on monitoring results and community feedback, ensuring that conservation goals remain responsive to emerging challenges.

7.3.8 Legal Reforms and Institutional Strengthening

Finally, to address governance fragmentation, the project should support legal reforms that clarify the mandates of relevant agencies, establish binding conservation obligations, and enhance the role of independent oversight bodies. A statutory Conservation Trust could be established to coordinate conservation funding, enforce compliance, and ensure transparency in project implementation.

By implementing this integrated mitigation strategy, the Great Nicobar Project can transform from a narrowly conceived infrastructural development into a model for inclusive, sustainable, and resilient development. This strategy not only safeguards biodiversity and cultural heritage but also empowers indigenous communities as stewards of their own futures, ensuring that conservation and development proceed hand in hand in one of the world's most ecologically and culturally significant regions.

Thus, the Great Nicobar project warrants far greater scrutiny in light of the island's exceptional ecological, archaeological, and cultural significance. As a living archive of prehistoric and historic trajectories, the island potentially preserves evidence of early human migrations across Southeast Asia, offering invaluable insights into the deep past of our species. It is home to the indigenous Nicobarese and the Shompen communities, whose unique cultural landscapes and lifeways have evolved in intimate dialogue with their environment.

Furthermore, recent discoveries of tool use by Nicobar long-tailed macaques (*Macaca fascicularis umbrosus*) position the island as a critical site for understanding human–primate behavioral evolution. Unlike more extensively studied mainland populations, the Nicobar macaques demonstrate how tool-use traditions may persist, fragment, or disappear under rapid ecological transformation, making their conservation a matter of preserving evolutionary processes, not simply species numbers (Figure 7.3). Biologically, Great Nicobar is a sanctuary for irreplaceable species, including the Nicobar megapode, and one of the world's few remaining nesting sites for the endangered leatherback turtle. Its rainforests, mangroves, and coral reefs are not only local keystones of biodiversity but also planetary assets – absorbing carbon, buffering climate change, and supporting vital oceanic corridors. The preservation of Great Nicobar is thus not merely a regional concern but a global responsibility, underscored by international frameworks such as the United Nations Declaration on

Figure 7.3 A Subadult Nicobar Long-Tailed Macaque (*Macaca fascicularis umbrosus*) Clings to the Bark of a Fallen Tree.

the Rights of Indigenous Peoples (UNDRIP), World Heritage conventions, and the Sustainable Development Goals. Protecting this island safeguards a shared ecological and evolutionary heritage for generations to come.

Bibliography

Abbaspoor, S., Rahman, K., Zinke, W., & Hoffman, K. L. (2023). Learning of object-in-context sequences in freely-moving macaques. *bioRxiv*, 2023.12.11.571113. https://doi.org/10.1101/2023.12.11.571113

Abegg, C., & Thierry, B. (2002). Macaque evolution and dispersal in insular south-east Asia: MACAQUE DISPERSAL IN INSULAR ASIA. *Biological Journal of the Linnean Society*, *75*(4), 555–576. https://doi.org/10.1046/j.1095-8312.2002.00045.x

Abney, K. (2023). *Infant-maternal behaviors during non-maternal infant handling interactions in captive long- tailed macaques* (Macaca fascicularis) [MS. Thesis, Central Washington University]. https://digitalcommons.cwu.edu/etd/1894

Agostini, I., & Visalberghi, E. (2005). Social influences on the acquisition of sex-typical foraging patterns by juveniles in a group of wild tufted capuchin monkeys (*Cebus nigritus*). *American Journal of Primatology*, *65*(4), 335–351. https://doi.org/10.1002/ajp.20120

Agrawal, H. N. (1969). An anthropological study of the Shompens of Great Nicobar. *Human Heredity*, *19*(3), 312–315. https://doi.org/10.1159/000152234

Al Hakim, R. R., Imtiyaaz, C. D., Setyawaty, D., Rahayu, F., & Rianti, P. (2022). Daily behaviour of long-tailed macaque in the captive, semi-wild, and wild habitats: Preliminary reports. *Indonesian Journal of Primatology*, *1*(1), 25–32. https://doi.org/10.29244/primatology.1.01.25-32

Alappatt, J. P. (2017). *Common forest plants of Andaman and Nicobar Islands*. Department of Environment and Forest, Andaman and Nicobar Administration.

Albanese, V., Berardi, R., Llorente, M., Huffman, M. A., & Leca, J.-B. (2022). Mechanisms of expression of object play: A comparative study of stone handling in two captive groups of long-tailed macaques (Macaca fascicularis fascicularis). *Behavioural Processes*, 104774. https://doi.org/10.1016/j.beproc.2022.104774

Albert, R. A. (2013). *The evolution of sex differences in tool use* [M.A. Thesis]. University of Missouri-Columbia.

Alcock, J. (1972). The evolution of the use of tools by feeding animals. *Evolution*, *26*(3), 464. https://doi.org/10.2307/2407020

Almécija, S., Hammond, A. S., Thompson, N. E., Pugh, K. D., Moyà-Solà, S., & Alba, D. M. (2021). Fossil apes and human evolution. *Science*, *372*(6542), eabb4363. https://doi.org/10.1126/science.abb4363

Altmann, J., & Samuels, A. (1992). Costs of maternal care: Infant-carrying in baboons. *Behavioral Ecology and Sociobiology*, *29*(6), 391–398. https://doi.org/10.1007/BF00170168

Ameneshewa, W., Kebede, Y., Unbushe, D., & Legesse, A. (2023). Indigenous knowledge and forest management practices among *Shekachoo* people in the *Sheka* Biosphere Reserve A case of *Shato* core area, South-west Ethiopia. *Cogent Social Sciences*, *9*(2), 2275937. https://doi.org/10.1080/23311886.2023.2275937

Amici, F., Kulik, L., Langos, D., & Widdig, A. (2019). Growing into adulthood-A review on sex differences in the development of sociality across macaques. *Behavioral Ecology and Sociobiology*, *73*(2), 18. https://doi.org/10.1007/s00265-018-2623-2

Anderson, D. G., Bissett, T. G., Yerka, S. J., Wells, J. J., Kansa, E. C., Kansa, S. W., Myers, K. N., DeMuth, R. C., & White, D. A. (2017). Sea-level rise and archaeological site destruction: An example from the southeastern United States using DINAA (Digital Index of North American Archaeology). *PLoS One*, *12*(11), e0188142. https://doi.org/10.1371/journal.pone.0188142

Anderson, J. R. (2011). A primatological perspective on death. *American Journal of Primatology*, *73*(5), 410–414. https://doi.org/10.1002/ajp.20922

Anderson, J. R., & Chamove, A. S. (1985). Early social experience and the development of self-aggression in monkeys. *Biology of Behaviour*, *10*, 147–157.

Andrews, P., Reynolds, S. C., & Bobe, R. (2022). Approaches to the Study of Past Environments. In S. C. Reynolds & R. Bobe (Eds.), *African Paleoecology and Human Evolution* (1st ed., pp. 7–14). Cambridge University Press. https://doi.org/10.1017/9781139696470.002

Angst, W. (1975). Basic Data and Concepts on the Social Organization of Macaca fascicularis. In *Primate Behavior* (pp. 325–388). Elsevier. https://doi.org/10.1016/B978-0-12-534004-5.50011-7

Anitha, K., Aneesh, A., Raghavan, R., Kanagavel, A., Augustine, T., & Joseph, S. (2013). Identifying Habitat Connectivity for Isolated Populations of Lion-Tailed Macaque (Macaca silenus) in Valparai Plateau, Western Ghats, India. *Primate Conservation*, *2013*(27), 91–97. https://doi.org/10.1896/052.027.0101

Anuar, S., & Ullah, K. (2011). Social Organization and Mating System of Macaca Fascicularis (Long Tailed Macaques). *International Journal of Biology*, *3*(2), p23. https://doi.org/10.5539/ijb.v3n2p23

Aplin, L. M., Farine, D. R., Morand-Ferron, J., & Sheldon, B. C. (2012). Social networks predict patch discovery in a wild population of songbirds. Proceedings of the Royal Society B: Biological Sciences, 279(1745), 4199–4205. https://doi.org/10.1098/rspb.2012.1591

Aplin, L. (2016). Understanding the multiple factors governing social learning and the diffusion of innovations. *Current Opinion in Behavioral Sciences*, *12*, 59–65. https://doi.org/10.1016/j.cobeha.2016.09.003

Archie, E. A., Altmann, J., & Alberts, S. C. (2014). Costs of reproduction in a long-lived female primate: Injury risk and wound healing. *Behavioral Ecology and Sociobiology*, *68*(7), 1183–1193. https://doi.org/10.1007/s00265-014-1729-4

Arlet, M. E., Balasubramaniam, K., Saha, R., Marty, P. R., Kaburu, S. S. K., Bliss-Moreau, E., Kaasik, A., Kodandaramaiah, U., & McCowan, B. (2020). Infant survival among free-living bonnet macaques (*Macaca radiata*) in South India. *International Journal of Primatology*.

Arora, K. (2010). Sustainable management of tropical forest through indigenous knowledge: A case study of Shompens of Great Nicobar Island. *Indian Journal of Traditional Knowledge*, *9*(3), 551–561.

Arora, K. (2018). *Indigenous forest management in the Andaman and Nicobar Islands, India*. Springer International Publishing. https://doi.org/10.1007/978-3-030-00033-2

Arre, A. M., Stumph, E., & Santos, L. R. (2021). Macaque species with varying social tolerance show no differences in understanding what other agents perceive. *Animal Cognition*. https://doi.org/10.1007/s10071-021-01485-5

Arroyo, A., Hirata, S., Matsuzawa, T., & de la Torre, I. (2016). Nut cracking tools used by captive chimpanzees (*Pan troglodytes*) and their comparison with early stone age percussive artefacts from Olduvai Gorge. *PLoS One*, *11*(11), e0166788. https://doi.org/10.1371/journal.pone.0166788

Arvind, L. S. (2024). Janki Andharia (Ed), Disaster studies: Exploring intersectionalities in the disaster discourse. *The International Journal of Community and Social Development*, *6*(3), 339–340. https://doi.org/10.1177/25166026241266664

August, B. (1840). *The Nicobar Islands or Frederiksoerne*. Royal Danish Library.

Aureli, F. (1992). Post-conflict behaviour among wild long-tailed macaques (Macaca fascicularis). *Behavioral Ecology and Sociobiology*, *31*(5), 329–337. https://doi.org/10.1007/BF00177773

Aureli, F., & Waal, F. B. M. de (Eds.). (2000). *Natural conflict resolution*. University of California Press.

Awaradi, S. A. (2022). Ethnographic hotspots of Andaman and Nicobar Islands: Urgency for multidisciplinary studies. *The Oriental Anthropologist: A Bi-Annual International Journal of the Science of Man*, *22*(1), 173–185. https://doi.org/10.1177/0972558X221110386

Baber, C. (2003). *Cognition and tool use: Forms of engagement in human and animal use of tools*. Taylor & Francis.

Babitz, M. A. (1999). *The effects of experience on tool use by Capuchin Monkeys* [PhD thesis]. University of St. Andres.

Ballesta, S., Reymond, G., Pozzobon, M., & Duhamel, J.-R. (2014). Compete to play: Trade-off with social contact in long-tailed macaques (*Macaca fascicularis*). *PLoS One*, *9*(12), e115965. https://doi.org/10.1371/journal.pone.0115965

Bandhu, S., Verma, S., & Sharma, V. N. (2024). *A study of Tribal tourism with special reference to India by using G.I.S. techniques*. https://doi.org/10.5281/ZENODO.12740096

Bandini, E., Grossmann, J., Funk, M., Albiach-Serrano, A., & Tennie, C. (2021). Naïve orangutans (*Pongo abelii* and *Pongo pygmaeus*) individually acquire nut-cracking using hammer tools. *American Journal of Primatology*, *83*(9). https://doi.org/10.1002/ajp.23304

Bandini, E., Harrison, R. A., & Motes-Rodrigo, A. (2022). Examining the suitability of extant primates as models of hominin stone tool culture. *Humanities and Social Sciences Communications*, *9*(1), 74. https://doi.org/10.1057/s41599-022-01091-x

Bandini, E., & Tennie, C. (2020). Exploring the role of individual learning in animal tool-use. *PeerJ*, *8*, e9877. https://doi.org/10.7717/peerj.9877

Bandopadhyay, P. C., & Carter, A. (2017a). Chapter 2 Introduction to the geography and geomorphology of the Andaman–Nicobar Islands. *Geological Society, London, Memoirs*, *47*(1), 9–18. https://doi.org/10.1144/M47.2

Bandopadhyay, P. C., & Carter, A. (2017b). Chapter 6 geological framework of the Andaman–Nicobar Islands. *Geological Society, London, Memoirs*, *47*(1), 75–93. https://doi.org/10.1144/M47.6

Bandopadhyay, P. C., & Carter, A. (2017c). Geological framework of the Andaman–Nicobar Islands. In P. C. Bandopadhyay & A. Carter (Eds.), *The Andaman–Nicobar accretionary ridge: Geology, tectonics and hazards* (Vol. 47, p. 0). Geological Society of London.

Bandopadhyay, P. C., & Carter, A. (2018). Paleogene tectonic and sedimentation history of the Andaman-Nicobar Accretionary Arc, Northeast Indian Ocean. In S. Bajpai, S. C. Tripathi, & V. Prasad (Eds.), *The Indian Paleogene* (pp. 91–112). Springer International Publishing. https://doi.org/10.1007/978-3-319-77443-5_2

Bandyopadhyay, R., & Yuwanond, P. (2018). Representation, resistance and cultural hybridity of the *Naga* Indigenous people in India. *Tourism Management Perspectives*, *26*, 164–171. https://doi.org/10.1016/j.tmp.2017.10.008

Banerjee, R. (2024). *Settler memory and Islanding: An archipelagic perspective* (PMML Occasional Paper, 2024) [Rehabilitation schemes (1965–80)]. Prime Ministers Museum and Library.

Barrett, B. J., McElreath, R. L., & Perry, S. E. (2017). Pay-off-biased social learning underlies the diffusion of novel extractive foraging traditions in a wild primate. *Proceedings of the Royal Society B: Biological Sciences*, *284*(1856), 20170358. https://doi.org/10.1098/rspb.2017.0358

Barrett, B. J., Monteza-Moreno, C. M., Dogandžić, T., Zwyns, N., Ibáñez, A., & Crofoot, M. C. (2018). Habitual stone-tool-aided extractive foraging in white-faced capuchins, *Cebus capucinus*. *Royal Society Open Science*, *5*(8), Article 8. https://doi.org/10.1098/rsos.181002

Barstow, M. (2018). IUCN red list of threatened species: Pterocarpus dalbergioides. *IUCN Red List of Threatened Species*. https://www.iucnredlist.org/en

Beck, B. B. (1980). *Animal tool behavior: The use and manufacture of tools by animals*. Garland STPM Press.

Bednekoff, P. A. (2015). Sentinel behavior: A review and prospectus. In *Advances in the study of behavior* (Vol. 47, pp. 115–145). Elsevier. https://doi.org/10.1016/bs.asb.2015.02.001

Begun, D. R. (2015). Fossil record of miocene hominoids. In W. Henke & I. Tattersall (Eds.), *Handbook of paleoanthropology* (pp. 1261–1332). Springer Berlin Heidelberg. https://doi.org/10.1007/978-3-642-39979-4_32

Behie, A. M., Teichroeb, J. A., & Malone, N. (Eds.). (2019). *Primate research and conservation in the anthropocene* (1st ed.). Cambridge University Press. https://doi.org/10.1017/9781316662021

Beier, P., & Noss, R. F. (1998). Do habitat corridors provide connectivity? *Conservation Biology*, *12*(6), 1241–1252. https://doi.org/10.1111/j.1523-1739.1998.98036.x

Bellard, C., Bertelsmeier, C., Leadley, P., Thuiller, W., & Courchamp, F. (2012). Impacts of climate change on the future of biodiversity. *Ecology Letters*, *15*(4), 365–377. https://doi.org/10.1111/j.1461-0248.2011.01736.x

Benito-Calvo, A., Carvalho, S., Arroyo, A., Matsuzawa, T., & de la Torre, I. (2015). First GIS analysis of modern stone tools used by wild chimpanzees (Pan troglodytes verus) in Bossou, Guinea, West Africa. *PLoS One*, *10*(3), e0121613. https://doi.org/10.1371/JOURNAL.PONE.0121613

Bennick, A. (2002). Interaction of plant polyphenols with saliva proteins. *Critical Reviews in Oral Biology & Medicine*, *13*(2), 184–196. https://doi.org/10.1177/154411130201300208

Bennis, K.L (Ed.). (2023). *Global volcanism program* (Report on Barren Island (India) - October 2023 No. 48:10; Bulletin of the Global Volcanism Network). https://volcano.si.edu/ShowReport.cfm?doi=10.5479/si.GVP.BGVN202310-260010

Bercovitch, F. B. (1991). Social stratification, social strategies, and reproductive success in primates. *Ethology and Sociobiology*, *12*(4), 315–333. https://doi.org/10.1016/0162-3095(91)90023-J

Bercovitch, F. B. (1995). Female cooperation, consortship maintenance, and male mating success in savanna baboons. *Animal Behaviour*, *50*(1), 137–149. https://doi.org/10.1006/anbe.1995.0227

Bercovitch, F. B. (2002). Sex-biased parental investment in primates. *International Journal of Primatology*, *23*(4), 905–921. https://doi.org/10.1023/A:1015585117114

Berenstain, L. (1986). Responses of long-tailed macaques to drought and fire in Eastern Borneo: A preliminary report. *Biotropica*, *18*(3), 257. https://doi.org/10.2307/2388494

Berman, C. M. (1982). The ontogeny of social relationships with group companions among free-ranging infant rhesus monkeys I. Social networks and differentiation. *Animal Behaviour*, *30*(1), 149–162. https://doi.org/10.1016/S0003-3472(82)80250-9

Bernstein, I. S. (1976). Dominance, aggression and reproduction in primate societies. *Journal of Theoretical Biology*, *60*(2), 459–472. https://doi.org/10.1016/0022-5193(76)90072-2

Bernstein, I. S. (2005). Macaque societies. A model for the study of social organization. *International Journal of Primatology*, *26*(5), 1199–1201. https://doi.org/10.1007/s10764-005-6467-z

Bezanson, M., & McNamara, A. (2019). The what and where of primate field research may be failing primate conservation. *Evolutionary Anthropology: Issues, News, and Reviews*. https://doi.org/10.1002/evan.21790

Bhan, S. (1992). Development of mother-infant relationship in rhesus monkeys (*Macaca mulatta*) in natural habitats. In P. K. Seth (Ed.), *Perspectives in primate biology* (Vol. 4, pp. 163–175). Today & Tomorrow's Printers and Publishers.

Bhandari, A., Kay, R. F., Williams, B. A., Tiwari, B. N., Bajpai, S., & Hieronymus, T. (2018). First record of the Miocene hominoid Sivapithecus from Kutch, Gujarat state, western India. *PLoS One*, *13*(11), e0206314. https://doi.org/10.1371/journal.pone.0206314

Bird, R. L. B., & Bird, D. W. (1997). Delayed reciprocity and tolerated theft: The behavioral ecology of food-sharing strategies. *Current Anthropology*, *38*(1), 49–78.

Biro, D., Haslam, M., & Rutz, C. (2013). Tool use as adaptation. *Philosophical Transactions of the Royal Society B: Biological Sciences*, *368*(1630), 20120408. https://doi.org/10.1098/rstb.2012.0408

Biro, D., Inoue-Nakamura, N., Tonooka, R., Yamakoshi, G., Sousa, C., & Matsuzawa, T. (2003). Cultural innovation and transmission of tool use in wild chimpanzees: Evidence from field experiments. *Animal Cognition*, *6*(4), 213–223. https://doi.org/10.1007/s10071-003-0183-x

Biswas, A. (2024). Tourism and Its Environmental Impact on the Andaman and Nicobar Islands. *International Journal for Research in Applied Science and Engineering Technology*, *12*(8), 128–131. https://doi.org/10.22214/ijraset.2024.63876

Bjorklund, D. F., & Gardiner, A. K. (2010). Object play and tool use: Developmental and evolutionary perspectives. In P. Nathan & A. D. Pellegrini (Eds.), *The Oxford handbook of the development of play*. Oxford University Press. https://doi.org/10.1093/oxfordhb/9780195393002.013.0013

Blersch, R., Beisner, B. A., Vandeleest, J. J., & McCowan, B. (2024). The dynamics of dominance in a 'despotic' society. *Animal Behaviour*, *218*, 55–64. https://doi.org/10.1016/j.anbehav.2024.09.009

Blurton Jones, N. G. (1987). Tolerated theft, suggestions about the ecology and evolution of sharing, hoarding and scrounging. *Social Science Information*, *26*(1), 31–54. https://doi.org/10.1177/053901887026001002

Boesch, C. (2012). *Wild cultures: A comparison between chimpanzee and human cultures*. Cambridge University Press.

Boesch, C. (2017). *Technical intelligence and culture: Nut cracking in humans and chimpanzees-Boesch-2017-American Journal of Physical Anthropology-Wiley Online Library*. https://onlinelibrary.wiley.com/doi/abs/10.1002/ajpa.23211

Boesch, C., & Boesch, H. (1989). Hunting behavior of wild chimpanzees in the Taï National Park. *American Journal of Physical Anthropology*, *78*(4), Article 4. https://doi.org/10.1002/ajpa.1330780410

Boesch, C., & Boesch, H. (1990). Tool use and tool making in wild Chimpanzees. *Folia Primatologica*, *54*(1–2), 86–99. https://doi.org/10.1159/000156428

Boesch, C., Head, J., & Robbins, M. M. (2009). Complex tool sets for honey extraction among chimpanzees in Loango National Park, Gabon. *Journal of Human Evolution*, *56*(6), 560–569. https://doi.org/10.1016/j.jhevol.2009.04.001

Boonkusol, D., Sanyathitiseree, P., Thongyuan, S., & Jangsuwan, N. (2018). Population and behavior surveys of long-tailed macaque (*Macaca fascicularis*) in the Old Town, Lopburi Province. *OnLine Journal of Biological Sciences*, *18*(2), 226–236. https://doi.org/10.3844/ojbsci.2018.226.236

Boose, K. J., White, F. J., & Meinelt, A. (2013). Sex differences in tool use acquisition in bonobos (*P an paniscus*): Bonobo sex differences in tool use. *American Journal of Primatology*, *75*(9), 917–926. https://doi.org/10.1002/ajp.22155

Borries, C., Sommer, V., & Srivastava, A. (1991). Dominance, age, and reproductive success in free-ranging female hanuman langurs (*Presbytis entellus*). *International Journal of Primatology*, *12*(3), 231–257. https://doi.org/10.1007/BF02547586

Borries, C., Sommer, V., & Srivastava, A. (1994). Weaving a tight social net: Allogrooming in free-ranging female langurs (*Presbytis entellus*). *International Journal of Primatology*, *15*(3), 421–443. https://doi.org/10.1007/BF02696102

Botting, J. (2021, December 11). *Long-tailed Macaque, Macaca fascicularis | New England primate conservancy*. https://neprimateconservancy.org/long-tailed-macaque/

Boussaoud, D. (2015). *Learning by observation in the macaque monkey under high experimental constraints*. Unpublished. https://doi.org/10.13140/RG.2.1.1387.4086

Bowles, S., & Hammerstein, P. (2023). A biological employment model of reproductive inequality. *Philosophical Transactions of the Royal Society B: Biological Sciences*, *378*(1883), 20220289. https://doi.org/10.1098/rstb.2022.0289

Bridger, R. (2022). *Great Nicobar development plan, Nicobar islands, India | EJAtlas*. Environmental Justice Atlas. https://ejatlas.org/conflict/great-nicobar-development-plan

Bril, B., Smaers, J., Steele, J., Rein, R., Nonaka, T., Dietrich, G., Biryukova, E., Hirata, S., & Roux, V. (2012). Functional mastery of percussive technology in nut-cracking and stone-flaking actions: Experimental comparison and implications for the evolution of the human brain. *Philosophical Transactions of the Royal Society B: Biological Sciences*, *367*(1585), 59–74. https://doi.org/10.1098/rstb.2011.0147

Bromham, L., Dinnage, R., Skirgård, H., Ritchie, A., Cardillo, M., Meakins, F., Greenhill, S., & Hua, X. (2021). Global predictors of language endangerment and the future of linguistic diversity. *Nature Ecology & Evolution*, *6*(2), 163–173. https://doi.org/10.1038/s41559-021-01604-y

Brotcorne, F., Giraud, G., Gunst, N., Fuentes, A., Wandia, I. N., Beudels-Jamar, R. C., Poncin, P., Huynen, M.-C., & Leca, J.-B. (2017). Intergroup variation in robbing and bartering by long-tailed macaques at Uluwatu Temple (Bali, Indonesia). *Primates*, *58*(4), 505–516. https://doi.org/10.1007/s10329-017-0611-1

Brown, C. R., Brown, M. B., & Shaffer, M. L. (1991). Food-sharing signals among socially foraging cliff swallows. *Animal Behaviour*, *42*(4), 551–564. https://doi.org/10.1016/S0003-3472(05)80239-8

Byrne, R. W., & Bates, L. A. (2010). Primate social cognition: Uniquely primate, uniquely social, or just unique? *Neuron*, *65*(6), 815–830. https://doi.org/10.1016/j.neuron.2010.03.010

Call, J., & Tennie, C. (2009). Animal culture: Chimpanzee table manners? *Current Biology*, *19*(21), R981–R983. https://doi.org/10.1016/j.cub.2009.09.036

Call, J., & Tomasello, M. (1994). The social learning of tool use by orangutans (*Pongo pygmaeus*). *Human Evolution*, *9*(4), 297–313. https://doi.org/10.1007/BF02435516

Cant, J. G. H. (1988). Positional behavior of long-tailed Macaques (*Macaca fascicularis*) in Northern Sumatra. *American Journal of Physical Anthropology*, *76*(1), 29–37. https://doi.org/10.1002/ajpa.1330760104

Canteloup, C., Cera, M. B., Barrett, B. J., & Van De Waal, E. (2021). Processing of novel food reveals payoff and rank-biased social learning in a wild primate. *Scientific Reports*, *11*(1), 9550. https://doi.org/10.1038/s41598-021-88857-6

Canteloup, C., Hoppitt, W., & Van De Waal, E. (2020). Wild primates copy higher-ranked individuals in a social transmission experiment. *Nature Communications*, *11*(1), 459. https://doi.org/10.1038/s41467-019-14209-8

Carne, C., Wiper, S., & Semple, S. (2011). Reciprocation and interchange of grooming, agonistic support, feeding tolerance, and aggression in semi-free-ranging Barbary macaques. *American Journal of Primatology*, *73*(11), 1127–1133. https://doi.org/10.1002/ajp.20979

Caro, T. (2007). Behavior and conservation: A bridge too far? *Trends in Ecology & Evolution*, *22*(8), 394–400. https://doi.org/10.1016/j.tree.2007.06.003

Carr, A., Ruhanen, L., & Whitford, M. (2016). Indigenous peoples and tourism: The challenges and opportunities for sustainable tourism. *Journal of Sustainable Tourism*, *24*(8–9), 1067–1079. https://doi.org/10.1080/09669582.2016.1206112

Carruthers, L., Ersek, V., Maher, D., Sanders, C., Tait, D., Soares, J., Floyd, M., Hashim, A. S., Helber, S., Garnett, M., East, H., Johnson, J. A., Ponta, G., & Sippo, J. Z. (2024). Sea-level rise and extreme Indian Ocean Dipole explain mangrove dieback in the Maldives. *Scientific Reports*, *14*(1), 27012. https://doi.org/10.1038/s41598-024-73776-z

Carvalho, S., & Almeida-Warren, K. (2019). Primate archaeology. In *Encyclopedia of Animal Behavior* (pp. 397–407). Elsevier. https://doi.org/10.1016/B978-0-12-809633-8.90156-0

Carvalho, S., Wessling, E. G., Abwe, E. E., Almeida-Warren, K., Arandjelovic, M., Boesch, C., Danquah, E., Diallo, M. S., Hobaiter, C., Hockings, K., Humle, T., Ikemeh, R. A., Kalan, A. K., Luncz, L., Ohashi, G., Pascual-Garrido, A., Piel, A., Samuni, L., Soiret, S., … Koops, K. (2022). Using nonhuman culture in conservation requires careful and concerted action. *Conservation Letters*, *n/a*(n/a), e12860. https://doi.org/10.1111/conl.12860

Casey, M. (2008, June 10). Monkeys that go fishing discovered. *NBC News*. https://www.nbcnews.com/id/wbna25078708

Cassini, M. H. (2020). Sexual size dimorphism and sexual selection in primates. *Mammal Review*, *50*(3), 231–239. https://doi.org/10.1111/mam.12191

Cawthon, L. K. A. (2006). *Primate factsheets: Long-tailed macaque (Macaca fascicularis) behavior*. Wisconsin National Primate Research Center. https://primate.wisc.edu/primate-info-net/pin-factsheets/pin-factsheet-long-tailed-macaque/

Cenni, C., Christie, J. B. A., Van der Pant, Y., Gunst, N., Vasey, P. L., Wandia, I. N., & Leca, J. (2022). Do monkeys use sex toys? Evidence of stone tool-assisted masturbation in free-ranging long-tailed macaques. *Ethology*, eth.13324. https://doi.org/10.1111/eth.13324

Cenni, C., Nord, C., Christie, J. B. A., Wandia, I. N., & Leca, J.-B. (2024). How does object play shape tool use emergence? Integrating observations and field experiments in longtailed macaques. *Animal Behaviour*, *218*, 239–254. https://doi.org/10.1016/j.anbehav.2024.09.001

Cenni, C., Thierry, B., Wandia, I. N., & Leca, J.-B. (2023). Tool-assisted water scooping in Balinese long-tailed macaques. *Behaviour*, *160*(8–9), 817–836. https://doi.org/10.1163/1568539X-bja10230

Chaimanee, Y., Chavasseau, O., Beard, K. C., Kyaw, A. A., Soe, A. N., Sein, C., Lazzari, V., Marivaux, L., Marandat, B., Swe, M., Rugbumrung, M., Lwin, T., Valentin, X., Zin-Maung-Maung-Thein, & Jaeger, J.-J. (2012). Late Middle Eocene primate from Myanmar and the initial anthropoid colonization of Africa. *Proceedings of the National Academy of Sciences*, *109*(26), 10293–10297. https://doi.org/10.1073/pnas.1200644109

Chamove, A. S., Anderson, J. R., & Nash, V. J. (1984). Social and environmental influences on self-aggression in monkeys. *Primates*, *25*(3), 319–325. https://doi.org/10.1007/BF02382270

Chandel, N. K. (1992). Dominance related visual attention toward adult male rhesus monkeys (*Macaca mulatta*) in free-ranging group. In P. K. Seth (Ed.), *Perspectives in primate biology* (Vol. 4, pp. 121–124). Today & Tomorro's Printers and Publishers.

Chandra, K., & Venkataraman, K. (2014). *473 publications 1,324 citations see profile*.

Chang, J. (2006). Indigenous tourism: The commodification and management of culture. *Annals of Tourism Research*, *33*(1), 282–283. https://doi.org/10.1016/j.annals.2005.10.003

Chapman, C. A., & Rothman, J. M. (2009). Within-species differences in primate social structure: Evolution of plasticity and phylogenetic constraints. *Primates*, *50*(1), 12–22. https://doi.org/10.1007/s10329-008-0123-0

Chatterjee, A. (2020). The science of the Andamans and the sign of the four: The distorted racial hierarchy of British imperial anthropology. *Shima: The International Journal of Research into Island Cultures*, *14*(2). https://doi.org/10.21463/shima.14.2.14

Chatterjee, M. (2003). *Social behaviour of wild bonnet macaques: Affiliative and agonistic behaviour of adult females* [Unpublished report]. Indian Academy of Sciences.

Chatterjee, M. (2012). *An Uncommon performing monkey: Behvioural flexibility and social diversity in the northern bonnet macaque* Macaca radiata radiata [Ph.D. thesis]. National Institute of Advanced Studies.

Cheng, J. T., Hemelrijk, C. K., Hentschel, T., Huchard, E., Kappeler, P. M., & Veldman, J. (2023). Editorial: Sex and gender effects on power, status, dominance, and leadership – An interdisciplinary look at human and other mammalian societies. *Frontiers in Ecology and Evolution*, *11*, 1340095. https://doi.org/10.3389/fevo.2023.1340095

Chin, Y. X. (2024). Biodiversity Record: Carrying of dead infant by long-tailed macaque. *Nature in Singapore*, *17*, 12. https://doi.org/10.26107/NIS-2024-0127

Chism, J., & Rogers, W. (1997). Male competition, mating success and female choice in a seasonally breeding primate (*Erythrocebus patas*). *Ethology*, *103*(2), 109–126. https://doi.org/10.1111/j.1439-0310.1997.tb00011.x

Choi, C. Q. (2009, November 11). *Human evolution: The orgin of tool use* [Health]. Live Science. https://www.livescience.com/7968-human-evolution-origin-tool.html

Chokkalingam, U. (2001). Secondary forest: A working definition and typology. *International Forestry Review*, *3*(1), 19–27.

Choudhury, A. (1988). Priority ratings for conservation of Indian primates. *Oryx*, *22*(2), 89–94. https://doi.org/10.1017/S0030605300027551

CIFOR-ICRAF. (2024). *About – Mangrove monitoring for climate change mitigation in India*. https://www.cifor-icraf.org/mangrove-monitoring-india/about/

Clancy, K. B. H., Hinde, K., & Rutherford, J. N. (Eds.). (2013). *Building babies: Primate development in proximate and ultimate perspective*. Springer.

Clutton-Brock, T. (2007). Sexual selection in males and females. *Science*, *318*(5858), 1882–1885. https://doi.org/10.1126/science.1133311

Clutton-Brock, T. H., & Huchard, E. (2013). Social competition and selection in males and females. *Philosophical Transactions of the Royal Society B: Biological Sciences*, *368*(1631), 20130074. https://doi.org/10.1098/rstb.2013.0074

Cnuts, D., & Rots, V. (2024). Examining the effect of post-depositional processes on the preservation and identification of stone tool residues from temperate environments: An experimental approach. *PLoS One*, *19*(10), e0309060. https://doi.org/10.1371/journal.pone.0309060

Coiner-Collier, S., Scott, R. S., Chalk-Wilayto, J., Cheyne, S. M., Constantino, P., Dominy, N. J., Elgart, A. A., Glowacka, H., Loyola, L. C., Ossi-Lupo, K., Raguet-Schofield, M., Talebi, M. G., Sala, E. A., Sieradzy, P., Taylor, A. B., Vinyard, C. J., Wright, B. W., Yamashita, N., Lucas, P. W., & Vogel, E. R. (2016). Primate dietary ecology in the context of food mechanical properties. *Journal of Human Evolution*, *98*, 103–118. https://doi.org/10.1016/j.jhevol.2016.07.005

Cooley, F. (2014). *The Shompen*. http://dice.missouri.edu/docs/asia-other/Shompen.pdf

Cooper, Z. (1993). Perceptions of time in the Andaman Islands. *World Archaeology*, *25*(2), 261–267. http://www.jstor.org/stable/124818

Cosby, A. E., Andriatsitohaina, B., Archie, R., Jules, C., McGoogan, K. C., Persram, M., Razafitsalama, M., Young, A., & Steffens, T. S. (2024). Primate conservation: A public issue? *American Journal of Primatology*, e23632. https://doi.org/10.1002/ajp.23632

Coussi-Korbel, S., & Fragaszy, D. M. (1995). On the relation between social dynamics and social learning. *Animal Behaviour*, *50*(6), 1441–1453. https://doi.org/10.1016/0003-3472(95)80001-8

Cowlishaw, G., & Dunbar, R. I. M. (1991). Dominance rank and mating success in male primates. *Animal Behaviour*, *41*(6), 1045–1056. https://doi.org/10.1016/S0003-3472(05)80642-6

Creighton, M. J. A., Greenberg, D. A., Reader, S. M., & Mooers, A. Ø. (2021). The role of behavioural flexibility in primate diversification. *Animal Behaviour*, *180*, 269–290. https://doi.org/10.1016/j.anbehav.2021.07.013

Das, S., David, R. C., Anand, A., Harikumar, S., Rajan, R., & Singh, M. (2020). Use of an embedded fruit by Nicobar Long-tailed Macaque *Macaca fascicularis umbrosus*: II. Demographic influences on choices of coconuts Cocos nucifera and pattern of forays to palm plantations. *Journal of Threatened Taxa*, *12*(11), 16407–16423. https://doi.org/10.11609/jott.6510.12.11.16407-16423

Das, S., Harpalani, M., Chaudhuri, B., Negi, R., Ali, S. T., & Singh, M. (2020). Use of an embedded fruit by the nicobar long-tailed macaque (*Macaca fascicularis umbrosus*): I. Familiarity to coconuts (*Cocos nucifera* L.) and temporal patterns of coconut foraging. *Primate Conservation*, *34*, 1–22.

Dasgupta, D., Banerjee, A., Dutta, A., Mitra, S., Banerjee, D., Karar, R., Karmakar, S., Bhattacharya, A., Ghosh, S., Bhattacharjee, P., & Paul, M. (2025). Decoding food solicitation techniques applied by free-ranging Hanuman langurs residing in an urban habitat. *Animal Cognition*, *28*(1), 25. https://doi.org/10.1007/s10071-024-01925-y

Davidson, I. (2016). Stone tools: Evidence of something in between culture and cumulative culture? In M. N. Haidle, N. J. Conard, & M. Bolus (Eds.), *Vertebrate paleobiology and paleoanthropology* (pp. 99–120). Springer. https://doi.org/10.1007/978-94-017-7426-0_10

Dawson, N. M., Coolsaet, B., Bhardwaj, A., Booker, F., Brown, D., Lliso, B., Loos, J., Martin, A., Oliva, M., Pascual, U., Sherpa, P., & Worsdell, T. (2024). Is it just conservation? A typology of Indigenous peoples' and local communities' roles in conserving biodiversity. *One Earth*, S2590332224002124. https://doi.org/10.1016/j.oneear.2024.05.001

de Jong, G., de Ruiter, J. R., & Haring, R. (1994). Genetic structure of a population with social structure and migration. *EXS*, *68*, 147–164. https://doi.org/10.1007/978-3-0348-8510-2_13

De Laat, E. J. A. M., Waasdorp, S., Roth, T. S., Massen, J. J. M., & Sterck, E. H. M. (2024). *Despotic long-tailed macaques benefit others in a group service paradigm*. https://doi.org/10.1101/2024.06.28.601131

De Ruiter, J. R., Van Hooff, J. A. R. A. M., & Scheffrahn, W. (1994). Social and genetic aspects of paternity in wild long-tailed macaques (*Macaca fascicularis*). *Behaviour*, *129*(3–4), 203–224. https://doi.org/10.1163/156853994X00613

de Waal, F. B. M. (1989). Food sharing and reciprocal obligations among chimpanzees. *Journal of Human Evolution*, *18*(5), 433–459. https://doi.org/10.1016/0047-2484(89)90074-2

de Waal, F. B. M. (2000). Attitudinal reciprocity in food sharing among brown capuchin monkeys. *Animal Behaviour*, *60*(2), Article 2. https://doi.org/10.1006/anbe.2000.1471

de Waal, F. B. M. (2007). *Chimpanzee politics: Power and sex among apes* (25. anniversary ed). Johns Hopkins Univ. Press.

de Beaune, S. A. (2004). The invention of technology: Prehistory and cognition. *Current Anthropology*, *45*(2), 139–162. https://doi.org/10.1086/381045

Debnath, H. S. (2004). *Mangroves of Andaman & Nicobar Islands: Taxonomy and ecology: A community profile*. Bishen Singh Mahendra Pal Singh.

Deborah Ramer, M. (2010, April 1). *The history of indigenous peoples and tourism | Cultural survival*. Cultural and Survival. https://www.culturalsurvival

.org/publications/cultural-survival-quarterly/history-indigenous-peoples-and-tourism

Declerck, M., Trifonova, N., Black, J., Hartley, J., & Scott, B. E. (2022). New strategic framework to structure Cumulative Impact Assessment (CIA). *International Marine Energy Journal*, *5*(3), 339–347. https://doi.org/10.36688/imej.5.339-347

Delson, E. (1993). Theropithecus fossils from Africa and India and the taxonomy of the genus. In N. G. Jablonski (Ed.), *Theropithecus: The rise and fall of a primate genus* (pp. 157–190). Cambridge University Press. https://doi.org/10.1017/CBO9780511565540.006

Desai, B. G. (2021). Ichnofabric analysis of bathyal chalks: The Miocene Inglis formation of the Andaman and Nicobar Islands, India. *Journal of Palaeogeography*, *10*(4), 494–508. https://doi.org/10.1016/j.jop.2021.11.002

Dileep, K., & Jose, B. (2014). Social behaviour in bonnet macaque (*Macaca Radiata*) with reference to grooming. *International Journal of Innovative Science, Engineering & Technology*, *1*(8), 340–343.

Diogo, R., Richmond, B. G., & Wood, B. (2012). Evolution and homologies of primate and modern human hand and forearm muscles, with notes on thumb movements and tool use. *Journal of Human Evolution*, *63*(1), 64–78. https://doi.org/10.1016/j.jhevol.2012.04.001

Director, Z.S. I. (Ed.) (with Rao, G. C., Subba Rao, N. V., Baqri, Q. H., Shastri, R. K., Haldar, B. P., Ghosh, & Mukhopadhyay, S. K.). (1995). *Bibliography on zoology of Andaman and Nicobar Islands (1845–1993)*. Zoological Survey of India.

Dixson, A. F. (2012). *Primate sexuality: Comparative studies of the prosimians, monkeys, apes, and humans*. Oxford University Press. https://doi.org/10.1093/acprof:osobl/9780199544646.001.0001

Dubreuil, L., & Nadel, D. (2015). The development of plant food processing in the Levant: Insights from use-wear analysis of Early Epipalaeolithic ground stone tools. *Philosophical Transactions of the Royal Society B: Biological Sciences*, *370*(1682), 20140357. https://doi.org/10.1098/rstb.2014.0357

Dunayer, E. S., & Berman, C. M. (2018). Infant handling among primates. *International Journal of Comparative Psychology*, *31*. https://doi.org/10.46867/ijcp.2018.31.02.06

Dunbar, R. I. M. (2012). Bridging the bonding gap: The transition from primates to humans. *Philosophical Transactions of the Royal Society B: Biological Sciences*, *367*(1597), 1837–1846. https://doi.org/10.1098/rstb.2011.0217

Dyble, M., Thompson, J., Smith, D., Salali, G. D., Chaudhary, N., Page, A. E., Vinicuis, L., Mace, R., & Migliano, A. B. (2016). Networks of food sharing reveal the functional significance of multilevel sociality in two hunter-gatherer groups. *Current Biology*, *26*(15), 2017–2021. https://doi.org/10.1016/j.cub.2016.05.064

Eichler Vercillo, U., Gonçalves Morato, R., de Almeida Cunha, A., de Marco, P., Strier, K. B., Mittermeier, R. A., & Luiz de Andrade Franco, J. (2022). Action plans for species conservation are an important tool to meet global and national biodiversity targets – A study case in Brazil. *Journal for Nature Conservation*, 126324. https://doi.org/10.1016/j.jnc.2022.126324

Elisabetta, V., Haslam, M., Spagnoletti, N., & Fragaszy, D. (2013). Use of stone hammer tools and anvils by bearded capuchin monkeys over time and space: Construction of an archeological record of tool use. *Journal of Archaeological Science*, *40*(8), Article 8. https://doi.org/10.1016/j.jas.2013.03.021

Ellwanger, A. L., & Lambert, J. E. (2018). Investigating niche construction in dynamic human-animal landscapes: Bridging ecological and evolutionary timescales. *International Journal of Primatology*. https://doi.org/10.1007/s10764-018-0033-y

Elton, S. (2002). A reappraisal of the locomotion and habitat preference of Theropithecus oswaldi. *Folia Primatologica; International Journal of Primatology*, *73*(5), 252–280. https://doi.org/10.1159/000067457

Engelhardt, A., Heistermann, M., Hodges, J. K., Nürnberg, P., & Niemitz, C. (2006). Determinants of male reproductive success in wild long-tailed macaques (*Macaca fascicularis*)-Male monopolisation, female mate choice or post-copulatory mechanisms? *Behavioral Ecology and Sociobiology*, *59*(6), 740–752. https://doi.org/10.1007/s00265-005-0104-x

Engelhardt, A., Hodges, J. K., Niemitz, C., & Heistermann, M. (2005). Female sexual behavior, but not sex skin swelling, reliably indicates the timing of the fertile phase in wild long-tailed macaques (Macaca fascicularis). *Hormones and Behavior*, *47*(2), 195–204. https://doi.org/10.1016/j.yhbeh.2004.09.007

EQUATIONS, INTACH Andaman & Nicobar Islands Chapter, Society for Andaman & Nicobar Ecology, Kalpavriksh, Jamsetji Tata Centre for Disaster Management (TISS), & ActionAid International India. (2008). *Rethink tourism in the Andamans – Towards building a base for sustainable tourism* [Summary research report.]. http://www.indiaenvironmentportal.org.in/files/Rethink%20Tourism%20in%20the%20Andamans.pdf

Etting, S. F., Isbell, L. A., & Grote, M. N. (2014). Factors increasing snake detection and perceived threat in captive rhesus macaques (*Macaca mulatta*). *American Journal of Primatology*, *76*(2), 135–145. https://doi.org/10.1002/ajp.22216

Fagen, R. (1982). Skill and flexibility in animal play behavior. *Behavioral and Brain Sciences*, *5*(1), 162–162. https://doi.org/10.1017/S0140525X00011006

Fairbanks, L. A. (1996). Individual differences in maternal style: Causes and consequences for mothers and offspring. In *Advances in the study of behavior* (Vol. 25, pp. 579–611). Academic Press. https://doi.org/10.1016/S0065-3454(08)60343-5

Falótico, T., Coutinho, P. H. M., Bueno, C. Q., Rufo, H. P., & Ottoni, E. B. (2018). Stone tool use by wild capuchin monkeys (*Sapajus libidinosus*) at Serra das Confusões National Park, Brazil. *Primates*. https://doi.org/10.1007/s10329-018-0660-0

Falótico, T., & Ottoni, E. (2015). Sexual bias in tool use by Capuchin Monkeys. *Folia Primatologica*, *86*(4), Article 4. https://doi.org/10.1159/000435825

Falótico, T., & Ottoni, E. B. (2014). Sexual bias in probe tool manufacture and use by wild bearded capuchin monkeys. *Behavioural Processes*, *108*, 117–122. https://doi.org/10.1016/j.beproc.2014.09.036

Falótico, T., Siqueira, J. O., & Ottoni, E. B. (2017). Digging up food: Excavation stone tool use by wild capuchin monkeys. *Scientific Reports*, *7*(1), 6278. https://doi.org/10.1038/s41598-017-06541-0

Falótico, T., Spagnoletti, N., Haslam, M., Luncz, L. V., Malaivijitnond, S., & Gumert, M. (2017). Analysis of sea almond (*Terminalia catappa*) cracking sites used by wild Burmese long-tailed macaques (*Macaca fascicularis aurea*). *American Journal of Primatology*, *79*(5), Article 5. https://doi.org/10.1002/ajp.22629

Feistner, A., & McGrew, W. C. (1989a). Food-sharing in primates: A critical review. In P. K. Seth & S. Seth (Eds.), *Prespectivesw in primate biology* (Vol. 3, pp. 21–36). Today & Tomorro's Printers and Publishers.

Feistner, A., & McGrew, W. C. (1989b). Food-sharing in primates: A critical review. In P. K. Seth & S. Seth (Eds.), *Prespectivesw in primate biology* (Vol. 3, pp. 21–36). Today & Tomorro's Printers and Publishers.

Fitch-Snyder, H., & Carter, J. (1993). Tool use to acquire drinking water by free-ranging lion-tailed macaques (Macaca silenus). *Laboratory Primate Newsletter*, *32*, 1–1

Ford, J. D., King, N., Galappaththi, E. K., Pearce, T., McDowell, G., & Harper, S. L. (2020). The resilience of indigenous peoples to environmental change. *One Earth*, *2*(6), 532–543. https://doi.org/10.1016/j.oneear.2020.05.014

Forest Survey of India. (2019). *India state of forest report 2019* (16th ed., Vol. 1). Forest Survey of India (Ministry of Environment Forest and Climate Change).

Fredman, T., & Whiten, A. (2008). Observational learning from tool using models by human-reared and mother-reared capuchin monkeys (Cebus apella). *Animal Cognition*, *11*(2), 295–309. https://doi.org/10.1007/s10071-007-0117-0

Fruth, B., & Hohmann, G. (2018). Food sharing across borders: First observation of intercommunity meat sharing by Bonobos at LuiKotale, DRC. *Human Nature*, *29*(2), 91–103. https://doi.org/10.1007/s12110-018-9311-9

Fuentes, A., Gumert, M. D., & Jones-Engel, L. (Eds.). (2011). *Monkeys on the edge: Ecology and management of long-tailed macaques and their interface with humans*. Cambridge University Press.

Funkhouser, J. A., Mayhew, J. A., Sheeran, L. K., Mulcahy, J. B., & Li, J.-H. (2018). Comparative investigations of social context-dependent dominance in captive chimpanzees (*Pan troglodytes*) and Wild Tibetan Macaques (*Macaca thibetana*). *Scientific Reports*, *8*(1). https://doi.org/10.1038/s41598-018-32243-2

Fürtbauer, I., Schülke, O., Heistermann, M., & Ostner, J. (2010). Reproductive and life history parameters of wild female macaca assamensis. *International Journal of Primatology*, *31*(4), 501–517. https://doi.org/10.1007/s10764-010-9409-3

Galef, B. G. (1992). The question of animal culture. *Human Nature*, *3*(2), 157–178. https://doi.org/10.1007/BF02692251

Galef, B. G., & Giraldeau, L.-A. (2001). Social influences on foraging in vertebrates: Causal mechanisms and adaptive functions. *Animal Behaviour*, *61*(1), 3–15. https://doi.org/10.1006/anbe.2000.1557

Gamalo, L. E., Ilham, K., Jones-Engel, L., Gill, M., Sweet, R., Aldrich, B., Phiapalath, P., Van Bang, T., Ahmed, T., Kite, S., Paramasivam, S., Seiha, H., Zainol, M. Z., Nielsen, D. R. K., Ruppert, N., Fuentes, A., & Hansen, M. F. (2024). Removal from the wild endangers the once widespread long-tailed macaque. *American Journal of Primatology*, *86*(3), e23547. https://doi.org/10.1002/ajp.23547

Garber, P. A. (1987). Foraging strategies among living primates. *Annual Review of Anthropology*, *16*, 339–364.

Georgiev, A. V., Muehlenbein, M. P., Prall, S. P., Emery Thompson, M., & Maestripieri, D. (2015). Male quality, dominance rank, and mating success in free-ranging rhesus macaques. *Behavioral Ecology*, *26*(3), 763–772. https://doi.org/10.1093/beheco/arv008

Getahun, D. A., Delson, E., & Seyoum, C. M. (2023). A review of *Theropithecus oswaldi* with the proposal of a new subspecies. *Journal of Human Evolution*, *180*, 103373. https://doi.org/10.1016/j.jhevol.2023.103373

Ghosh, B., Bandyopadhyay, D., & Morishita, T. (2017). Chapter 7 Andaman–Nicobar Ophiolites, India: Origin, evolution and emplacement. *Geological Society, London, Memoirs*, *47*(1), 95–110. https://doi.org/10.1144/M47.7

Gilbert, C. C., Ortiz, A., Pugh, K. D., Campisano, C. J., Patel, B. A., Singh, N. P., Fleagle, J. G., & Patnaik, R. (2020). New middle miocene ape (Primates: Hylobatidae) from Ramnagar, India fills major gaps in the hominoid fossil record. *Proceedings of the Royal Society B: Biological Sciences*, *287*(1934), 20201655. https://doi.org/10.1098/rspb.2020.1655

Gilbert, C. C., Ortiz, A., Pugh, K. D., Campisano, C. J., Patel, B. A., Singh, N. P., Fleagle, J. G., & Patnaik, R. (2025). Additional analyses of stem catarrhine and hominoid dental morphology support Kapi ramnagarensis as a stem hylobatid. *Journal of Human Evolution*, *199*, 103628. https://doi.org/10.1016/j.jhevol.2024.103628

Gilbert, C. C., Patel, B. A., Singh, N. P., Campisano, C. J., Fleagle, J. G., Rust, K. L., & Patnaik, R. (2017). New sivaladapid primate from Lower Siwalik deposits surrounding Ramnagar (Jammu and Kashmir State), India. *Journal of Human Evolution*, *102*, 21–41. https://doi.org/10.1016/j.jhevol.2016.10.001

Gingerich, P. D. (1984). Primate evolution: Evidence from the fossil record, comparative morphology, and molecular biology. *American Journal of Physical Anthropology*, *27*(S5), 57–72. https://doi.org/10.1002/ajpa.1330270504

Girard-Buttoz, C., Heistermann, M., Rahmi, E., Agil, M., Fauzan, P. A., & Engelhardt, A. (2014a). Costs of and investment in mate-guarding in wild long-tailed macaques (Macaca fascicularis): Influences of female characteristics and male–female social bonds. *International Journal of Primatology*, *35*(3–4), 701–724. https://doi.org/10.1007/s10764-014-9775-3

Girard-Buttoz, C., Heistermann, M., Rahmi, E., Marzec, A., Agil, M., Fauzan, P. A., & Engelhardt, A. (2014b). Mate-guarding constrains feeding activity but not energetic status of wild male long-tailed macaques (*Macaca fascicularis*). *Behavioral Ecology and Sociobiology*, *68*(4), 583–595. https://doi.org/10.1007/s00265-013-1673-8

Global Volcanism Program. (2025). Barren Island (260001). Smithsonian Institution. https://volcano.si.edu/volcano.cfm?vn=260001

Goldstone, L. G., Sommer, V., Nurmi, N., Stephens, C., & Fruth, B. (2016). Food begging and sharing in wild bonobos (*Pan paniscus*): Assessing relationship quality? *Primates*, *57*(3), Article 3. https://doi.org/10.1007/s10329-016-0522-6

Gomez, L., Joseph, T., Heinrich, S., Wright, B., & D'Cruze, N. (2023). Illegal trade of pangolins in India with international trade links: An analysis of seizures from 1991 to 2022. *European Journal of Wildlife Research*, *69*(4), 85. https://doi.org/10.1007/s10344-023-01708-9

Gonçalves, A., & Carvalho, S. (2019). Death among primates: A critical review of non-human primate interactions towards their dead and dying. *Biological Reviews*, brv.12512. https://doi.org/10.1111/brv.12512

Govt. of India. (1956). *Andaman and Nicobar_Gazette_June_30_1956 PAT regulation*. https://andaman.gov.in/admin-pannel/pressupload/1-2-AN_Gazette_june_30_1956%20PAT%20Regulation.pdf

Govt. of India. (2018). *Historical background of Andaman | District South Andaman, Government of Andaman and Nicobar | India*. https://southandaman.nic.in/history/

Govt. of India. (2020). *Andaman UT administration*. https://andaman.gov.in/about

Govt. of India. (2022a). *History | District Nicobar, Government of Andaman and Nicobar | India*. https://nicobars.andaman.nic.in/history/

Govt. of India. (2022b). *The official website of ministry of environment, forest and climate change, Government of India*. https://moef.gov.in/en/

Goutham-Bharathi, M.P., Roy, S. Dam, Krishnan, P., Kaliyamoorthy, M. and Immanuel, Titus. (2014) "Species diversity and distribution of mangroves in Andaman and Nicobar Islands, India" *Botanica Marina*, vol. 57, no. 6, pp. 421–432. https://doi.org/10.1515/bot-2014-0033

Grampp, M., Sueur, C., Van De Waal, E., & Botting, J. (2019). Social attention biases in juvenile wild vervet monkeys: Implications for socialisation and social learning processes. *Primates*, *60*(3), 261–275. https://doi.org/10.1007/s10329-019-00721-4

Green, S. M. (1981). Sex differences and age gradations in vocalizations of Japanese and Lion-tailed monkeys (*Macaca fuscata and Macaca silenus*). *American Zoologist*, *21*(1), 165–183. https://doi.org/10.1093/icb/21.1.165

Gruber, T., Luncz, L., Mörchen, J., Schuppli, C., Kendal, R. L., & Hockings, K. (2019). Cultural change in animals: A flexible behavioural adaptation to human disturbance. *Palgrave Communications*, *5*(1), 1–9. https://doi.org/10.1057/s41599-019-0271-4

Gruber, T., Muller, M. N., Reynolds, V., Wrangham, R., & Zuberbühler, K. (2011). Community-specific evaluation of tool affordances in wild chimpanzees. *Scientific Reports*, *1*(1). https://doi.org/10.1038/srep00128

Grueter, C. C., Robbins, A. M., Abavandimwe, D., Vecellio, V., Ndagijimana, F., Ortmann, S., Stoinski, T. S., & Robbins, M. M. (2016). Causes, mechanisms, and consequences of contest competition among female mountain gorillas in Rwanda. *Behavioral Ecology*, *27*(3), 766–776. https://doi.org/10.1093/beheco/arv212

Gumert, M. D. (2007). Grooming and infant handling interchange in Macaca fascicularis: The relationship between infant supply and grooming payment. *International Journal of Primatology*, *28*(5), 1059–1074. https://doi.org/10.1007/s10764-007-9202-0

Gumert, M. D. (2010). Dominance and reciprocity in the grooming relationships of female long-tailed macaques (Macaca fascicularis) in Indonesia. In S. Gursky & J. Supriatna (Eds.), *Indonesian primates* (pp. 309–341). Springer. https://doi.org/10.1007/978-1-4419-1560-3_18

Gumert, M. D., Hamada, Y., & Malaivijitnond, S. (2013). Human activity negatively affects stone tool-using Burmese long-tailed macaques *Macaca fascicularis aurea* in Laem Son National Park, Thailand. *Oryx*, *47*(4), 535–543. https://doi.org/10.1017/S0030605312000130

Gumert, M. D., Hoong, L. K., & Malaivijitnond, S. (2011). Sex differences in the stone tool-use behavior of a wild population of burmese long-tailed macaques (*Macaca fascicularis aurea*). *American Journal of Primatology*, *73*(12), 1239–1249. https://doi.org/10.1002/ajp.20996

Gumert, M. D., Kluck, M., & Malaivijitnond, S. (2009). The physical characteristics and usage patterns of stone axe and pounding hammers used by long-tailed macaques in the Andaman Sea region of Thailand. *American Journal of Primatology*, *71*(7), 594–608. https://doi.org/10.1002/ajp.20694

Gumert, M. D., & Malaivijitnond, S. (2012). Marine prey processed with stone tools by Burmese long-tailed macaques (Macaca fascicularis aurea) in intertidal habitats. *American Journal of Physical Anthropology*, *149*(3), 447–457. https://doi.org/10.1002/ajpa.22143

Gumert, M. D., & Malaivijitnond, S. (2013). Long-tailed macaques select mass of stone tools according to food type. *Philosophical Transactions of the Royal Society B: Biological Sciences*, *368*(1630), 20120413–20120413. https://doi.org/10.1098/rstb.2012.0413

Gunst, N., Huffman, M., & Leca, J.-B. (2007). Japanese macaque cultures: Inter- and intra-troop behavioural variability of stone handling patterns across 10 troops. *Behaviour*, *144*(3), 251–281. https://doi.org/10.1163/156853907780425712

Hadi, I., Suryobroto, B., & Perwitasari-Farajallah, D. (2007). Food preference of semi-provisioned macaques based on feeding duration and foraging party size. *HAYATI Journal of Biosciences*, *14*(1), Article 1. https://doi.org/10.4308/hjb.14.1.13

Hall, K. R. L., & Schaller, G. B. (1964). Tool-using behavior of the California sea otter. *Journal of Mammalogy*, *45*(2), 287–298. https://doi.org/10.2307/1376994

Hambali, K., Ismail, A., & Md-Zain, B. M. (2012). *Daily activity budget of long-tailed macaques (Macaca fascicularis) in Kuala Selangor Nature Park*, *12*(4).

Hammer, R., Stribos, M. S., Boehm, P. M., Pink, K. E., Herzele, J., Wallner, B., Huffman, M. A., Massen, J. J. M., & Pflüger, L. S. (2023). A novel methodological approach for group classification during fission of a semi-free-ranging group of Japanese macaques (*Macaca fuscata*). *American Journal of Primatology*. https://doi.org/10.1002/ajp.23463

Hansen, M. F., Ang, A., Fuentes, A., & Nijman, V. (2024). IUCN red list of threatened species: Macaca fascicularis. *IUCN Red List of Threatened Species, e.T12551A273015436*. https://www.iucnredlist.org/en

Harmand, S., Lewis, J. E., Feibel, C. S., Lepre, C. J., Prat, S., Lenoble, A., Boës, X., Quinn, R. L., Brenet, M., Arroyo, A., Taylor, N., Clément, S., Daver, G., Brugal, J.-P., Leakey, L., Mortlock, R. A., Wright, J. D., Lokorodi, S., Kirwa, C., … Roche, H. (2015). 3.3-million-year-old stone tools from Lomekwi 3, West Turkana, Kenya. *Nature*, *521*(7552), 310–315. https://doi.org/10.1038/nature14464

Hasan, M. U., Widayati, K. A., Tsuji, Y., & Rianti, P. (2023). Feeding ecology of free-ranging long-tailed macaques in East Java, Indonesia: Relationship with human food availability. *Primates*, *64*(4), 429–438. https://doi.org/10.1007/s10329-023-01062-z

Haslam, M. (2012). Towards a prehistory of primates. *Antiquity*, *86*(332), Article 332. https://doi.org/10.1017/S0003598X00062773

Haslam, M. (2018). Food hardness and stone tool weight in wild primate nut-cracking. *bioRxiv*. https://doi.org/10.1101/267542

Haslam, M. (2024). Primate archaeology 3.1. *American Journal of Biological Anthropology*, *184*(1), e24919. https://doi.org/10.1002/ajpa.24919

Haslam, M., Falótico, T., & Luncz, L. (2018). Recognizing culture in wild primate tool use. In L. D. Di Paolo, F. Di Vincenzo, & De PetrilloF. (Eds.), *Evolution of primate social cognition* (pp. 199–209). Springer International Publishing. https://doi.org/10.1007/978-3-319-93776-2_13

Haslam, M., Fujii, J., Espinosa, S., Mayer, K., Ralls, K., Tinker, M. T., & Uomini, N. (2019). Wild sea otter mussel pounding leaves archaeological traces. *Scientific Reports*, *9*(1). https://doi.org/10.1038/s41598-019-39902-y

Haslam, M., Gumert, M. D., Biro, D., Carvalho, S., & Malaivijitnond, S. (2013). Use-wear patterns on wild macaque stone tools reveal their behavioural history. *PLoS One*, *8*(8), e72872. https://doi.org/10.1371/journal.pone.0072872

Haslam, M., Hernandez-Aguilar, A., Ling, V., Carvalho, S., de la Torre, I., DeStefano, A., Du, A., Hardy, B., Harris, J., Marchant, L., Matsuzawa, T., McGrew, W., Mercader, J., Mora, R., Petraglia, M., Roche, H., Visalberghi, E., & Warren, R. (2009). Primate archaeology. *Nature*, *460*(7253), Article 7253. https://doi.org/10.1038/nature08188

Haslam, M., Hernandez-Aguilar, R. A., Proffitt, T., Arroyo, A., Falótico, T., Fragaszy, D., Gumert, M., Harris, J. W. K., Huffman, M. A., Kalan, A. K., Malaivijitnond, S., Matsuzawa, T., McGrew, W., Ottoni, E. B., Pascual-Garrido, A., Piel, A., Pruetz, J., Schuppli, C., Stewart, F., … Luncz, L. V. (2017). Primate archaeology evolves. *Nature Ecology & Evolution*, *1*(10), Article 10. https://doi.org/10.1038/s41559-017-0286-4

Haslam, M., Luncz, L., Pascual-Garrido, A., Falótico, T., Malaivijitnond, S., & Gumert, M. (2016). Archaeological excavation of wild macaque stone tools. *Journal of Human Evolution*, *96*, 134–138. https://doi.org/10.1016/j.jhevol.2016.05.002

Haslam, M., Pascual-Garrido, A., Malaivijitnond, S., & Gumert, M. (2016). Stone tool transport by wild Burmese long-tailed macaques (*Macaca fascicularis aurea*). *Journal of Archaeological Science: Reports*, *7*, 408–413. https://doi.org/10.1016/j.jasrep.2016.05.040

Haunhorst, C. B., Fürtbauer, I., Schülke, O., & Ostner, J. (2020). Female macaques compete for 'power' and 'commitment' in their male partners. *Evolution and Human Behavior*, *41*(2), 117–125. https://doi.org/10.1016/j.evolhumbehav.2019.11.001

Hector, A. C. K., Seyfarth, R. M., & Raleigh, M. J. (1989). Male parental care, female choice and the effect of an audience in vervet monkeys. *Animal Behaviour*, *38*(2), 262–271. https://doi.org/10.1016/S0003-3472(89)80088-0

Henazi, S. P., & Barrett, L. (1999). The value of grooming to female primates. *Primates*, *40*(1), 47–59. https://doi.org/10.1007/BF02557701

Henrich, J. (2009). The evolution of innovation-enhancing institutions. In M. J. O'Brien & S. J. Shennan (Eds.), *Innovation in cultural systems* (pp. 99–120). The MIT Press. https://doi.org/10.7551/mitpress/8102.003.0012

Hill, C. M., & Webber, A. D. (2010). Perceptions of nonhuman primates in human-wildlife conflict scenarios. *American Journal of Primatology*, *72*(10), 919–924. https://doi.org/10.1002/ajp.20845

Hiller, M. A., Jarvis, B. C., Lisa, H., Paulson, L. J., Pollard, E. H. B., & Stanley, S. A. (2004). Recent trends in illegal logging and a brief discussion of their causes: A case study from Gunung Palung National Park, Indonesia. *Journal of Sustainable Forestry*, *19*(1–3), 181–212. https://doi.org/10.1300/J091v19n01_09

Hinde, K., Power, M. L., & Oftedal, O. T. (2009). Rhesus Macaque milk: Magnitude, sources, and consequences of individual variation over lactation. *American Journal of Physical Anthropology*, *138*(2), 148–157. https://doi.org/10.1002/ajpa.20911

Hiscock, P. (1985). The need for a taphonomic perspective in stone artefact analysis. *Queensland Archaeological Research*, 2, 82–97. https://doi.org/10.25120/qar.2.1985.197

Hobaiter, C., Poisot, T., Zuberbühler, K., Hoppitt, W., & Gruber, T. (2014). Social network analysis shows direct evidence for social transmission of tool use in wild Chimpanzees. *PLoS Biology*, *12*(9), e1001960.

Hockings, K. J., Bryson-Morrison, N., Carvalho, S., Fujisawa, M., Humle, T., McGrew, W. C., Nakamura, M., Ohashi, G., Yamanashi, Y., Yamakoshi, G., & Matsuzawa, T. (2015). Tools to tipple: Ethanol ingestion by wild chimpanzees using leaf-sponges. *Royal Society Open Science*, *2*(6), 150150. https://doi.org/10.1098/rsos.150150

Hoegh-Guldberg, O., Mumby, P. J., Hooten, A. J., Steneck, R. S., Greenfield, P., Gomez, E., Harvell, C. D., Sale, P. F., Edwards, A. J., Caldeira, K., Knowlton, N., Eakin, C. M., Iglesias-Prieto, R., Muthiga, N., Bradbury, R. H., Dubi, A.,

& Hatziolos, M. E. (2007). Coral reefs under rapid climate change and ocean acidification. *Science*, *318*(5857), 1737–1742. https://doi.org/10.1126/science.1152509

Hohmann, G. (1988). A case of simple tool use in wild liontailed macaques (*Macaca silenus*). *Primates*, *29*(4), 565–567. https://doi.org/10.1007/BF02381143

Hohmann, G. (1991). Comparative analyses of age- and sex-specific patterns of vocal behaviour in four species of old world monkeys. *Folia Primatologica*, *56*(3), 133–156. https://doi.org/10.1159/000156538

Hohmann, G. (2009). The diets of non-human primates: Frugivory, food processing, and food sharing. In J.-J. Hublin & M. P. Richards (Eds.), *The evolution of hominin diets* (pp. 1–14). Springer Netherlands. https://doi.org/10.1007/978-1-4020-9699-0_1

Honess, P. E. (2021). Behavioral biology of Macaques | 30 | Behavioral biology of laboratory. In *Behavioral biology of laboratory animals* (1st ed., p. 38). CRC Press, Taylor & Francis Group. https://www.taylorfrancis.com/chapters/edit/10.1201/9780429019517-30/behavioral-biology-macaques-paul-honess?context=ubx&refId=a5a01d79-0564-49b8-89ca-137ab9c92c43

Hrdy, S. (1981). *'Nepotists' and 'altruists': The behavior of old females among Macaques and Langur Monkeys (1981)* (pp. 59–76). Stanford University Press.

Huffman, M. A. (1996). Acquisition of innovative cultural behaviors in nonhuman primates: A case study of stone handling, a socially transmitted behavior in Japanese Macaques. In *Social learning in animals* (pp. 267–289). Elsevier. https://doi.org/10.1016/B978-012273965-1/50014-5

Huffman, M. A., Leca, J.-B., & Nahallage, C. A. D. (2010). Cultured Japanese Macaques: A multidisciplinary approach to stone handling behavior and its implications for the evolution of behavioral tradition in Nonhuman Primates. In N. Nakagawa, NakamichiM. , & H. Sugiura (Eds.), *The Japanese Macaques* (Vol. 0, pp. 191–219). Springer Japan. https://doi.org/10.1007/978-4-431-53886-8_9

Huffman, M. A., & Quiatt, D. (1986). Stone handling by Japanese macaques (*Macaca fuscata*): Implications for tool use of stone. *Primates*, *27*(4), 413–423. https://doi.org/10.1007/BF02381887

Hughes, J. K., Elton, S., & O'Regan, H. J. (2008). Theropithecus and 'Out of Africa' dispersal in the Plio-Pleistocene. *Journal of Human Evolution*, *54*(1), 43–77. https://doi.org/10.1016/j.jhevol.2007.06.004

Hussain, A., Kausar, T., Siddique, T., Kabir, K., An, Q. U., Rukhsar, F., Gorsi, F. I., Yaqub, S., Kauser, S., Rehman, A., Najam, A., Haroon, H., Rafiu, A., Korma, S. A., & Mahdi, A. A. (2024). Physiological and biochemical variations of naturally ripened mango (Mangifera Indica L.) with synthetic calcium carbide and ethylene. *Scientific Reports*, *14*(1), 2121. https://doi.org/10.1038/s41598-024-52483-9

Iki, S., & Kutsukake, N. (2023). Play face in Japanese macaques reflects the sender's play motivation. *Animal Cognition*, *26*(3), 849–859. https://doi.org/10.1007/s10071-022-01730-5.

International Survival. (2024). *The Shompen*. https://www.survivalinternational.org/tribes/shompen

ISFR. (2019). Andaman and Nicobar. In *India state forest report* (Vol. ii).

Jadeja, S. J., Gole, S. S., Apte, D. A., & Jabestin, A. (2016). *First nesting record of leatherback sea turtles on the west coast of Galathea Bay, Great Nicobar island, after the 2004 Indian ocean tsunami with notes on nest predation. 23.*

Jaeggi, A. V., & Gurven, M. (2013). Natural cooperators: Food sharing in humans and other primates: Natural cooperators: Food sharing in humans and other primates. *Evolutionary Anthropology: Issues, News, and Reviews*, *22*(4), Article 4. https://doi.org/10.1002/evan.21364

Jaeggi, A. V., Stevens, J. M. G., & van Schaik, C. P. (2010). Tolerant food sharing and reciprocity is precluded by despotism among bonobos but not chimpanzees. *American Journal of Physical Anthropology*, *143*(1), 41–51. https://doi.org/10.1002/ajpa.21288

Jaeggi, A. V., & Van Schaik, C. P. (2011). The evolution of food sharing in primates. *Behavioral Ecology and Sociobiology*, *65*(11), 2125–2140. https://doi.org/10.1007/s00265-011-1221-3

Jain, P. (2015, February 3). *Conservation and Sustainable development of Island Ecosystems: Challenges in Indian context - Academike*. https://www.lawctopus.com/academike/conservation-sustainable-development-island-ecosystems-challenges-indian-context/

Joly, M., Micheletta, J., De Marco, A., Langermans, J. A., Sterck, E. H. M., & Waller, B. M. (2017). Comparing physical and social cognitive skills in macaque species with different degrees of social tolerance. *Proceedings of the Royal Society B: Biological Sciences*, *284*(1862), 20162738. https://doi.org/10.1098/rspb.2016.2738

Kaburu, S. S. K., Beisner, B., Balasubramaniam, K. N., Marty, P. R., Bliss-Moreau, E., Mohan, L., Rattan, S. K., Arlet, M. E., Atwill, E. R., & McCowan, B. (2019). Interactions with humans impose time constraints on urban-dwelling rhesus macaques (Macaca mulatta). *Behaviour*, 1–28. https://doi.org/10.1163/1568539X-00003565

Kalbitzer, U., & Jack, K. M. (Eds.). (2018). Primate life histories, sex roles, and adaptability: Essays in honour of Linda M. Fedigan. Springer International Publishing. https://doi.org/10.1007/978-3-319-98285-4

Kandhari Malhotra, S. (2024). *Empowering indigenous communities: A vision for sustainable tourism - ET HospitalityWorld*. ETHospitalityWorld.Com. https://hospitality.economictimes.indiatimes.com/news/speaking-heads/empowering-indigenous-communities-a-vision-for-sustainable-tourism/107528125

Kappeler, P. M. (2002). Sexual selection in primates: New and comparative perspectives. *Evolutionary Anthropology: Issues, News, and Reviews*, *11*(5), 173–175. https://doi.org/10.1002/evan.10023

Kappeler, P. M., & Van Schaik, C. P. (2002). Evolution of primate social systems. *International Journal of Primatology*, *23*(4), 707–740. https://doi.org/10.1023/A:1015520830318

Kar, A., Mazumdar, A., & Dey, S. (2016). Water resources management challenges in the Andaman and Nicobar Islands. *Integrated and Sustainable Water Management: Science and Technology*, 168–177. https://doi.org/10.17491/cgsi/2016/95963

Kar, R. C. (2019). *Negritos' Andaman forest and civilized world*. Anjali Kar.

Karayathil, A. K., Gurusiddappa, L. H., & Shivasundari, R. S. M. (2023). Human-Monkey (Macaca Radiata) Conflict in Chamundi Hill-Mysuru, Karnataka. *World Journal of Environmental Biosciences*, *12*(2), 19–25. https://doi.org/10.51847/dpGQL4yAXY

Kassim, N., Hambali, K., & Amir, A. (2017). Nutritional composition of fruits selected by long-tailed macaques (Macaca fascicularis) in Kuala Selangor, Malaysia. *Tropical Life Sciences Research*, *28*(1), 91–101. https://doi.org/10.21315/tlsr2017.28.1.6

Kawai, M. (1965). Newly-acquired pre-cultural behavior of the natural troop of Japanese Monkeys on Koshima Islet. *Primates*, *6*(1), 1–30.

Kay, R. F. (2018). 100 years of primate paleontology. *American Journal of Physical Anthropology*, *165*(4), 652–676. https://doi.org/10.1002/ajpa.23429

Kazahari, N., & Agetsuma, N. (2010). Mechanisms determining relationships between feeding group size and foraging success in food patch use by Japanese macaques (*Macaca fuscata*). *Behaviour*, *147*(11), 1481–1500.

Kennedy, K. A. R., & Ciochon, R. L. (1999). A canine tooth from the Siwaliks: First recorded discovery of a fossil ape? *Human Evolution*, *14*(3), 231–253. https://doi.org/10.1007/BF02440159

Keskinen, M., & Kummu, M. (2010). *Impact assessment in the Mekong – Review of Strategic Environmental Assessment (SEA) & Cumulative Impact Assessment (CIA)*. Aalto University.

Kik, A., Adamec, M., Aikhenvald, A. Y., Bajzekova, J., Baro, N., Bowern, C., Colwell, R. K., Drozd, P., Duda, P., Ibalim, S., Jorge, L. R., Mogina, J., Ruli, B., Sam, K., Sarvasy, H., Saulei, S., Weiblen, G. D., Zrzavy, J., & Novotny, V. (2021). Language and ethnobiological skills decline precipitously in Papua New Guinea, the world's most linguistically diverse nation. *Proceedings of the National Academy of Sciences of the United States of America*, *118*(22), e2100096118. https://doi.org/10.1073/pnas.2100096118

Kokko, H., & Jennions, M. D. (2008). Parental investment, sexual selection and sex ratios. *Journal of Evolutionary Biology*, *21*(4), 919–948. https://doi.org/10.1111/j.1420-9101.2008.01540.x

Kononowicz, T. W., Rolando, F., Maigre, L., Sirigu, A., Duhamel, J.-R., Ballesta, S., & Wirth, S. (2025). Spatio-temporal dynamics of ingroup interactions in macaques. *Scientific Reports*, *15*, 30913. https://doi.org/10.1038/s41598-025-16391-w

Koops, K. (2020). Chimpanzee termite fishing etiquette. *Nature Human Behaviour*. https://doi.org/10.1038/s41562-020-0895-9

Krishna, B. A., Singh, M., & Singh, M. (2006). Population dynamics of a group of lion-tailed macaques *(Macaca silenus)* inhabiting a rainforest fragment in the Western Ghats, India. *Folia Primatologica*, *77*(5), 377–386. https://doi.org/10.1159/000093703

Krishnakumar, M. V. (2009). Development or despoilation? The Andaman Islands under colonial and postcolonial regimes. Shima: *The International Journal of Research into Island Cultures*, 3(2), 104–117

Kuběnová, B., Ostner, J., Schülke, O., Majolo, B., Šmilauer, P., & Konečná, M. (2019). The effect of dominance rank on the distribution of different types of male–infant–male interactions in Barbary Macaques (*Macaca sylvanus*). *International Journal of Primatology*, *40*(3), 300–315. https://doi.org/10.1007/s10764-019-00086-x

Kulik, L., Amici, F., Langos, D., & Widdig, A. (2015). Sex differences in the development of social relationships in rhesus macaques (*Macaca mulatta*). *International Journal of Primatology*, *36*(2), 353–376. https://doi.org/10.1007/s10764-015-9826-4

Kumar, N., Yadav, B. P., Tyagi, A., & Jaswal, A. K. (2012). Trend and spatial distribution of rainfall & rainy days over Andaman & Nicobar Islands. *Natural Hazards*, *63*(2), 575–587. https://doi.org/10.1007/s11069-012-0173-x

Kumar, R., Korff, M., & Sudhir, K. (2018). *A gender Atlas of India: With scorecard*. SAGE Publications, Inc. https://doi.org/10.4135/9789353287832

Kumar, S. (1985). The people of Nancowry Island: Their changing condition. *Human Science*, *34*, 134–145.

Kumara, H. N., Kumar, A., & Singh, M. (2020). *Macaca fascicularis ssp. umbrosa: The IUCN red list of threatened species 2021* [Data set]. https://doi.org/10.2305/IUCN.UK.2021-1.RLTS.T39791A17985345.en

Kummer, H., & Goodall, J. (1985). Conditions of innovative behaviour in primates. *Philosophical Transactions of the Royal Society B: Biological Sciences*, *308*(1135), Article 1135. https://doi.org/10.1098/rstb.1985.0020

Laínez, J. M. L. (2016). View of use of leaf-wrapping as a feeding technique by captive white-faced capuchin monkeys (*Cebus capucinus*) at the "Rosy Walther" metropolitan zoo, Honduras. *Neotropical Primates*, *23*(1), 26–29.

Lambert, J. E. (2013). Evolutionary biology of ape and monkey feeding and nutrition. In W. Henke & I. Tattersall (Eds.), *Handbook of paleoanthropology* (pp. 1–27). Springer https://doi.org/10.1007/978-3-642-27800-6_39-3

Laura, E. H., Matthew, W. T., Caley, M. O, Biren A. P., & Zeresenay A. (2026). Did modern human carpal morphology evolve from knuckle walking traits? Proc Biol Sci 1 May 2026; *293* (2071): 20260556. https://doi.org/10.1098/rspb.2026.055

Leca, J. B., Huffman, M. A., & Gunst, N. (2012). Thirty years of stone handling tradition in Arashiyama macaques: Implications for cumulative culture and tool use in non-human primates. In *The Monkeys of Stormy Mountain: 60 Years of Primatological Research on the Japanese Macaques of Arashiyama* (pp. 223–257). Cambridge University Press.

Leca, J.-B., Gunst, N., Gardiner, M., & Wandia, I. N. (2021). Acquisition of object-robbing and object/food-bartering behaviours: A culturally maintained token economy in free-ranging long-tailed macaques. *Philosophical Transactions of the Royal Society B: Biological Sciences*, *376*(1819), 20190677. https://doi.org/10.1098/rstb.2019.0677

Leca, J.-B., Gunst, N., & Huffman, M. A. (2010). The first case of dental flossing by a Japanese macaque (Macaca fuscata): Implications for the determinants of behavioral innovation and the constraints on social transmission. *Primates*, *51*(1), 13–22. https://doi.org/10.1007/s10329-009-0159-9

Lemorini, C., Plummer, T. W., Braun, D. R., Crittenden, A. N., Ditchfield, P. W., Bishop, L. C., Hertel, F., Oliver, J. S., Marlowe, F. W., Schoeninger, M. J., & Potts, R. (2014). Old stones' song: Use-wear experiments and analysis of the Oldowan quartz and quartzite assemblage from Kanjera South (Kenya). *Journal of Human Evolution*, *72*, 10–25. https://doi.org/10.1016/j.jhevol.2014.03.002

Lewis, G. E. (1937). Taxonomic syllabus of Siwalik fossil anthropoids. *American Journal of Science*, *s5-34*(200), 139–147. https://doi.org/10.2475/ajs.s5-34.200.139

Li, B., Li, W., Liu, C., Yang, P., & Li, J. (2022). Diverse diets and low-fiber, low-tannin foraging preferences: Foraging criteria of Tibetan macaques (Macaca thibetana) at low altitude in Huangshan. *Ecology and Evolution*, *12*(10), e9338. https://doi.org/10.1002/ece3.9338

Li, Y., Ma, G., Zhou, Q., & Huang, Z. (2020). Ranging patterns and foraging patch utilization of Assamese macaques inhabiting limestone forests in southwest Guangxi, China. *Global Ecology and Conservation*, *21*, e00816. https://doi.org/10.1016/j.gecco.2019.e00816

Lobo, I. A. (2021, September 24). *The threatened future of the Jarawa Tribe*. The Borgen Project. https://borgenproject.org/jarawa-tribe/

Lonsdorf, E. V. (2005). Sex differences in the development of termite-fishing skills in the wild chimpanzees, *Pan troglodytes schweinfurthii*, of Gombe National Park, Tanzania. *Animal Behaviour*, *70*(3), Article 3. https://doi.org/10.1016/j.anbehav.2004.12.014

Lucas, P. W., & Corlett, R. T. (1991). Relationship between the diet of Macaca fascicularis and forest phenology. *Folia Primatologica*, *57*(4), 201–215. https://doi.org/10.1159/000156587

Lucas, P. W., & Corlett, R. T. (1998). Seed dispersal by long-tailed macaques. *American Journal of Primatology*, *45*(1), 29–44. https://doi.org/10.1002/(SICI)1098-2345(1998)45:1%253C29::AID-AJP4%253E3.0.CO;2-Y

Luetjens, C. M., Fuchs, A., Baker, A., & Weinbauer, G. F. (2020). Group size experiences with enhanced pre- and postnatal development studies in the long-tailed macaque (*Macaca fascicularis*). *Primate Biology*, *7*(1), 1–4. https://doi.org/10.5194/pb-7-1-2020

Luncz, L. V., Braun, D. R., Marreiros, J., Bamford, M., Zeng, C., Pacome, S. S., Junghenn, P., Buckley, Z., Yao, X., & Carvalho, S. (2022). Chimpanzee wooden tool analysis advances the identification of percussive technology. *iScience*, *25*(11), 105315. https://doi.org/10.1016/j.isci.2022.105315

Luncz, L. V., Gill, M., Svensson, M. S., Kulik, L., & Malaivijitnond, S. (2019). Group-specific archaeological signatures of stone tool use in wild macaques. *eLife*, *8*, 1–21. https://doi.org/10.7554/eLife.46961

Luncz, L. V., Sirianni, G., Mundry, R., & Boesch, C. (2018). Costly culture: Differences in nut-cracking efficiency between wild chimpanzee groups. *Animal Behaviour*, *137*, 63–73. https://doi.org/10.1016/j.anbehav.2017.12.017

Luncz, L. V., Slania, N. E., Almeida-Warren, K., Carvalho, S., Falótico, T., Malaivijitnond, S., Arroyo, A., De La Torre, I., & Proffitt, T. (2024). Tool skill impacts the archaeological evidence across technological primates. *Scientific Reports*, *14*(1), 16556. https://doi.org/10.1038/s41598-024-67048-z

Luncz, L. V., Tan, A., Haslam, M., Kulik, L., Proffitt, T., Malaivijitnond, S., & Gumert, M. (2017). Resource depletion through primate stone technology. *eLife*, *6*. https://doi.org/10.7554/eLife.23647

Luncz, L. V., Wittig, R. M., & Boesch, C. (2015). Primate archaeology reveals cultural transmission in wild chimpanzees (*Pan troglodytes verus*). *Philosophical Transactions of the Royal Society B: Biological Sciences*, *370*(1682), 20140348. https://doi.org/10.1098/rstb.2014.0348

Macdonald, S. (2024). *The unintended impacts of tourism on coral reefs*. Women in Ocean Science. https://www.womeninoceanscience.com/blog/2020/5/26/the-unintended-impacts-of-tourism-on-coral-reefs

Macellini, S., Maranesi, M., Bonini, L., Simone, L., Rozzi, S., Ferrari, P. F., & Fogassi, L. (2012). Individual and social learning processes involved in the acquisition and generalization of tool use in macaques. *Philosophical Transactions of the Royal Society B: Biological Sciences*, *367*(1585), 24–36. https://doi.org/10.1098/rstb.2011.0125

Machanda, Z. P., & Rosati, A. G. (2020). Shifting sociality during primate ageing. *Philosophical Transactions of the Royal Society B: Biological Sciences*, *375*(1811), 20190620. https://doi.org/10.1098/rstb.2019.0620

MacKinnon, K. C., & Fuentes, A. (2011). Primates, niche construction, and social complexity: The roles of social cooperation and altruism. In R. W. Sussman & C. R. Cloninger (Eds.), *Origins of altruism and cooperation* (pp. 121–143). Springer. https://doi.org/10.1007/978-1-4419-9520-9_7

Macnaughtan, D. (2013). *The Shompen People of Great Nicobar Island, Indian Ocean* (pp. 1–9).

Maestripieri, D. (1994). Mother-infant relationships in three species of macaques (Macaca mulatta, M. nemestrina, M. arctoides). II. The social environment. *Behaviour*, *131*(1/2), 97–113.

Maestripieri, D. (1996). Social communication among captive stump-tailed macaques (Macaca arctoides). *International Journal of Primatology*, *17*(5), 785–802. https://doi.org/10.1007/BF02735264

Maestripieri, D. (2005). Gestural communication in three species of macaques (*Macaca mulatta*, *M. nemestrina*, *M. arctoides*): Use of signals in relation to dominance and social context. *Gesture*, *5*(1–2), 57–73. https://doi.org/10.1075/gest.5.1.06mae

Maestripieri, D., Hoffman, C. L., Anderson, G. M., Carter, C. S., & Higley, J. D. (2008). Mother–infant interactions in free-ranging rhesus macaques: Relationships between physiological and behavioral variables. *Physiology & Behavior*, *96*(0), 613. https://doi.org/10.1016/j.physbeh.2008.12.016

Maestripieri, D., & Wallen, K. (1997). Affiliative and submissive communication in rhesus macaques. *Primates*, *38*(2), 127–138. https://doi.org/10.1007/BF02382003

Malaivijitnond, S., Lekprayoon, C., Tandavanittj, N., Panha, S., Cheewatham, C., & Hamada, Y. (2007). Stone-tool usage by Thai long-tailed macaques (*Macaca fascicularis*). *American Journal of Primatology*, *69*(2), 227–233. https://doi.org/10.1002/ajp.20342

Maldonado-Celis, M. E., Yahia, E. M., Bedoya, R., Landázuri, P., Loango, N., Aguillón, J., Restrepo, B., & Guerrero Ospina, J. C. (2019). Chemical composition of Mango (Mangifera indica L.) Fruit: Nutritional and phytochemical compounds. *Frontiers in Plant Science*, *10*, 1073. https://doi.org/10.3389/fpls.2019.01073

Malik, J. N., Johnson, F. C., Khan, A., Sahoo, S., Irshad, R., Paul, D., Arora, S., Baghel, P. K., & Chopra, S. (2019). Tsunami records of the last 8000 years in the Andaman Island, India, from mega and large earthquakes: Insights on recurrence interval. *Scientific Reports*, *9*(1), 18463. https://doi.org/10.1038/s41598-019-54750-6

Malik, I. H. (2024) Can political ecology be decolonised? A dialogue with Paul Robbins. Geo: Geography and Environment, 11, e00140. Available from: https://doi.org/10.1002/geo2.140

Mallapur, A. (2013). Macaque tourism: Implications for their management and conservation. In S. Radhakrishna, M. A. Huffman, & A. Sinha (Eds.), *The macaque connection* (pp. 93–105). Springer. https://doi.org/10.1007/978-1-4614-3967-7_6

Manson, J. H. (1994). Mating patterns, mate choice, and birth season heterosexual relationships in free-ranging rhesus macaques. *Primates*, *35*(4), 417–433. https://doi.org/10.1007/BF02381951

Marchant, L. F. (1992). Chimpanzees, tools, and termites: Hand preference or handedness? *Current Anthropology*, *33*(1), Article 1. https://doi.org/10.1086/204041

Maréchal, L., Semple, S., Majolo, B., & MacLarnon, A. (2016). Assessing the effects of tourist provisioning on the health of wild barbary macaques in Morocco. *PLoS One*, *11*(5), e0155920. https://doi.org/10.1371/journal.pone.0155920

Markham, A. C., & Gesquiere, L. R. (2017). Costs and benefits of group living in primates: An energetic perspective. *Philosophical Transactions of the Royal Society B: Biological Sciences*, *372*(1727), 20160239. https://doi.org/10.1098/rstb.2016.0239

Marzke, M. W. (2013). Tool making, hand morphology and fossil hominins. *Philosophical Transactions of the Royal Society B: Biological Sciences*, *368*(1630), 20120414. https://doi.org/10.1098/rstb.2012.0414

Massen, J. J. M., Luyten, I. J. A. F., Spruijt, B. M., & Sterck, E. H. M. (2011). Benefiting friends or dominants: Prosocial choices mainly depend on rank position in long-tailed macaques (*Macaca fascicularis*). *Primates*, *52*(3), 237–247. https://doi.org/10.1007/s10329-011-0244-8

Matsuda, I., Clauss, M., Tuuga, A., Sugau, J., Hanya, G., Yumoto, T., Bernard, H., & Hummel, J. (2017). Factors affecting leaf selection by foregut-fermenting Proboscis Monkeys: New insight from in vitro digestibility and toughness of leaves. *Scientific Reports*, *7*(1), 42774. https://doi.org/10.1038/srep42774

Matsumura, S. (1999). The evolution of "egalitarian" and "despotic" social systems among macaques. *Primates*, *40*(1), 23–31. https://doi.org/10.1007/BF02557699

Mayhew, J. A., Funkhouser, J. A., & Wright, K. R. (2020). Considering social play in primates: A case study in Juvenile Tibetan Macaques (*Macaca thibetana*). In J.-H. Li, L. Sun, & P. M. Kappeler (Eds.), *The behavioral ecology of the Tibetan Macaque* (pp. 93–117). Springer International Publishing. https://doi.org/10.1007/978-3-030-27920-2_6

Mazumder, J. (2018). *Field notes on Nicobar long-tailed macaques from Great Nicobar, Katchall and Little Nicobar*. Unpublished Field notes

Mazumder, J. (2022). *Investigating diverse factors affecting tool use in monkeys through a study of wild Nicobar long-tailed macaques (Macaca fascicularis umbrosus) in India and captive rhesus macaques (*Macaca mulatta*) in the USA* [PhD thesis]. IISER Mohali.

Mazumder, J., & Chauhan, P. R. (2025). Revisiting "tool" for a more unified and holistic definition in animal behavior. *Evolutionary Anthropology: Issues, News, and Reviews*, *34*(3), 1–10. https://doi.org/10.1002/evan.70014

Mazumder, J., & Kaburu, S. S. K. (2020). Object manipulation and tool use in nicobar long-tailed macaques (*Macaca fascicularis umbrosus*). *International Journal of Primatology*, *41*, 141–159. https://doi.org/10.1007/s10764-020-00141-y

Mazumder, J., & Kaburu, S. S. K. (2021). First report of food sharing among nicobar long-tailed macaques. *Quaternary International*, *603*, 31–39. https://doi.org/10.1016/j.quaint.2020.11.049

McGrew, W. C., Tutin, C. E. G., & Wrangham, R. W. (2001). Charting cultural variation in chimpanzees. Behaviour, *138*(11-12), 1481–1516.

Melin, A. D., Young, H. C., Mosdossy, K. N., & Fedigan, L. M. (2014). Seasonality, extractive foraging and the evolution of primate sensorimotor intelligence. *Journal of Human Evolution*, *71*, 77–86. https://doi.org/10.1016/j.jhevol.2014.02.009

Menon, G., & Krishnaswamy, M. N. (2024, March 1). Watch | Why were ex-servicemen moved to Nicobar in 1970s? *The Hindu*. https://www.thehindu.com/news/national/watch-why-were-ex-servicemen-moved-to-nicobar-in-1970s/article67902937.ece

Mercader, J. (2002). Excavation of a Chimpanzee stone tool site in the African Rainforest. *Science*, *296*(5572), 1452–1455. https://doi.org/10.1126/science.1070268

Mercader, J., Barton, H., Gillespie, J., Harris, J., Kuhn, S., Tyler, R., & Boesch, C. (2007). 4,300-year-old chimpanzee sites and the origins of percussive stone technology. *Proceedings of the National Academy of Sciences*, *104*(9), 3043–3048. https://doi.org/10.1073/pnas.0607909104

Meunier, M., Monfardini, E., & Boussaoud, D. (2007). Learning by observation in rhesus monkeys. *Neurobiology of Learning and Memory*, *88*(2), 243–248. https://doi.org/10.1016/j.nlm.2007.04.015

Micheletta, J., Waller, B. M., Panggur, M. R., Neumann, C., Duboscq, J., Agil, M., & Engelhardt, A. (2012). Social bonds affect anti-predator behaviour in a tolerant species of macaque, *Macaca nigra*. *Proceedings of the Royal Society B: Biological Sciences*, *279*(1744), 4042–4050. https://doi.org/10.1098/rspb.2012.1470

Milich, K. M., & Maestripieri, D. (2016). Sex or power? The function of male displays in rhesus macaques. *Behaviour*, *153*(3), 245–261. https://doi.org/10.1163/1568539X-00003340

Mishra, P. S., Kumara, H. N., Thiyagesan, K., Singh, M., Velankar, A. D., & Pal, A. (2020). Chaos in coexistence: Perceptions of farmers towards long-tailed macaques (*Macaca fascicularis umbrosus*) Related to Crop Loss on Great Nicobar Island. *Primate Conservation*, *34*, 175–183.

Mishra, P. S., Pal, A., Velankar, A. D., Kumara, H. N., & Singh, M. (2020). Do males bond? A study of male-male relationships in Nicobar long-tailed macaques *Macaca fascicularis umbrosus*. *Journal of Biosciences*, *45*(1), 22. https://doi.org/10.1007/s12038-020-9995-y

Mishra, P. S., Pal, A., Velankar, A. D., Kumara, H. N., Singh, M., & Cooper, M. (2020). Does rank rule? Rank-related grooming patterns in Nicobar long-tailed macaques *Macaca fascicularis umbrosus*. *Primates*. https://doi.org/10.1007/s10329-020-00807-4

Mishra, R. K., Singh, P. P., Rai, N., Desai, S., Pandey, P., Tiwary, S. K., Tamang, R., Suravajhala, P., Shrivastava, P., Thangaraj, K., van Driem, G., & Chaubey, G. (2024). Reconstructing the population history of the Nicobarese. *European Journal of Human Genetics*, 1–7. https://doi.org/10.1038/s41431-024-01720-w

Mitani, J. C., Call, J., Kappeler, P. M., Palombit, R. A., & Silk, J. B. (Eds.). (2012). The Evolution of Primate Societies. University of Chicago Press. https://press.uchicago.edu/ucp/books/book/chicago/E/bo12986137.html

Mosley, H., & Haslam, M. (2016). Extending material cognition to primate tool use. *Quaternary International*, *405*, 70–77. https://doi.org/10.1016/j.quaint.2015.11.020

Murty, C. V. R., Jain, S. K., Sheth, A. R., Jaiswal, A., & Dash, S. R. (2006). Response and recovery in India after the December 2004 great Sumatra Earthquake and Indian Ocean Tsunami. *Earthquake Spectra*, *22*(S3), 731–758. https://doi.org/10.1193/1.2206137

Nahallage, C. A. D., Leca, J.-B., & Huffman, M. A. (2016). Stone handling, an object play behaviour in macaques: Welfare and neurological health implications of a bio-culturally driven tradition. *Behaviour*, *153*(6–7), 845–869. https://doi.org/10.1163/1568539X-00003361

Narasimmarajan, K., & Raghunathan, C. (2012). Status of long tailed macaque (Macaca fascicularis umbrosa) and Conservation of the recovery population in Great Nicobar Island, India. *Wildlife Biology in Practice*, *8*(2), Article 2. https://doi.org/10.2461/wbp.2012.8.6

Newman, L. E., Testard, C., DeCasien, A. R., Chiou, K. L., Watowich, M. M., Janiak, M. C., Pavez-Fox, M. A., Rosado, M. R. S., Cooper, E. B., Costa, C. E., Petersen, R. M., Montague, M. J., Platt, M. L., Brent, L. J. N., Snyder-Mackler, N., & Higham, J. P. (2023). The biology of aging in a social world: Insights from free-ranging rhesus macaques. *Neuroscience and Biobehavioral Reviews*, *154*, 105424. https://doi.org/10.1016/j.neubiorev.2023.105424

OECD. (2019). *The Illegal Wildlife Trade in Southeast Asia: Institutional capacities in Indonesia, Singapore, Thailand and Viet Nam*. OECD. https://doi.org/10.1787/14fe3297-en

O'Malley, R. C., & Fedigan, L. M. (2005). Evaluating social influences on food-processing behavior in white-faced capuchins (*Cebus capucinus*). *American Journal of Physical Anthropology*, *127*(4), 481–491. https://doi.org/10.1002/ajpa.20095

Oren, G., Shapira, A., Lifshitz, R., Vinepinsky, E., Cohen, R., Fried, T., Hadad, G. P., & Omer, D. (2024). Vocal labeling of others by nonhuman primates. *Science*, *385*(6712), 996–1003. https://doi.org/10.1126/science.adp3757

Ostner, J., & Schülke, O. (2018). Linking sociality to fitness in primates: A call for mechanisms. In M. Naguib, J. C. Mitani, L. W. Simmons, L. Barrett, S. D. Healy, & M. Zuk (Eds.), *Advances in the study of behavior* (Vol. 50, pp. 127–175). Elsevier. https://doi.org/10.1016/bs.asb.2017.12.001

Otani, Y., Bernard, H., Wong, A., Tangah, J., Tuuga, A., Hanya, G., & Matsuda, I. (2020). Factors influencing riverine utilization patterns in two sympatric macaques. *Scientific Reports*, *10*(1), 15749. https://doi.org/10.1038/s41598-020-72606-2

Overduin-de Vries, A. M., Spruijt, B. M., & Sterck, E. H. M. (2014). Long-tailed macaques (*Macaca fascicularis*) understand what conspecifics can see in a competitive situation. *Animal Cognition*, *17*(1), 77–84. https://doi.org/10.1007/s10071-013-0639-6

Overduin-de Vries, A. M., Vermande, M. M., Hessen, D. J., & Sterck, E. H. M. (2024). The ability to inhibit impulses is related to social behavior in long-tailed macaques. *American Journal of Primatology*, *86*(2), e23587. https://doi.org/10.1002/ajp.23587

Owen, A. D., & Hanley, N. (Eds.). (2004). *The economics of climate change*. Routledge.

Pal, A. (2018). *A study on reproductive behaviour of Nicobar long-tailed macaque (*Macaca fascicularis umbrosus*) in Nicobar Islands, India.* [Ph.D. Thesis]. Manipal University.

Pal, A., Kumara, H. N., Mishra, P. S., Velankar, A. D., & Singh, M. (2017). Extractive foraging and tool-aided behaviors in the wild Nicobar long-tailed macaque (Macaca fascicularis umbrosus). *Primates*. https://doi.org/10.1007/s10329-017-0635-6

Pal, A., Kumara, H. N., Mishra, P. S., Velankar, A. D., & Singh, M. (2018). Between-group encounters in Nicobar long-tailed macaque (*Macaca fascicularis umbrosus*). *Ethology Ecology & Evolution*, *30*(6), 582–599. https://doi.org/10.1080/03949370.2018.1459866

Pal, A., Kumara, H. N., Velankar, A. D., Mishra, P. S., & Singh, M. (2018). Demography and birth seasonality in the Nicobar long-tailed macaque (*Macaca fascicularis umbrosus*). *Current Science*, *114*(8), 1732–1737. https://doi.org/10.18520/cs/v114/i08/1732-1737

Palombit, R. A. (1992). A preliminary study of vocal communication in wild long-tailed macaques (Macaca fascicularis). I. Vocal repertoire and call emission. *International Journal of Primatology*, *13*(2), 143–182. https://doi.org/10.1007/BF02547839

Parr, L. A., & Heintz, M. (2009). Facial expression recognition in rhesus monkeys, Macaca mulatta. *Animal Behaviour*, *77*(6), 1507–1513. https://doi.org/10.1016/j.anbehav.2009.02.024

Pascual-Garrido, A., Carvalho, S., & Almeida-Warren, K. (2023). Primate archaeology 3.0. *American Journal of Biological Anthropology*, *n/a*(n/a). https://doi.org/10.1002/ajpa.24835

Patnaik, R., Cerling, T. E., Uno, K. T., & Fleagle, J. G. (2014). Diet and habitat of siwalik primates *Indopithecus, Sivaladapis* and *Theropithecus*. *Annales Zoologici Fennici*, *51*(1–2), 123–142. https://doi.org/10.5735/086.051.0214

Paukner, A., & Suomi, S. J. (2008). Sex differences in play behavior in juvenile tufted capuchin monkeys (Cebus apella). *Primates*, *49*(4), 288–291. https://doi.org/10.1007/s10329-008-0095-0

Pavelka, M. S. M. (1990). Do old female monkeys have a specific social role? *Primates*, *31*(3), 363–373. https://doi.org/10.1007/BF02381107

Pelletier, A. N., Kaufman, T., Mohak, S., Milan, R., Nahallage, C. A. D., Huffman, M. A., Gunst, N., Rompis, A., Wandia, I. N., Putra, I. G. A. A., Pellis, S. M., & Leca, J.-B. (2017). Behavior systems approach to object play: Stone handling repertoire as a measure of propensity for complex foraging and percussive tool use in the Genus *Macaca*. *Animal Behaviour and Cognition*, *4*(4), 455–473. https://doi.org/10.26451/abc.04.04.05.2017

Pellis, S. M., Pellis, V. C., Pelletier, A., & Leca, J.-B. (2019). Is play a behavior system, and, if so, what kind? *Behavioural Processes*, *160*, 1–9. https://doi.org/10.1016/j.beproc.2018.12.011

Perry, S. E., Barrett, B. J., & Godoy, I. (2017). Older, sociable capuchins (*Cebus capucinus*) invent more social behaviors, but younger monkeys innovate more in other contexts. *Proceedings of the National Academy of Sciences*, *114*(30), 7806–7813. https://doi.org/10.1073/pnas.1620739114

Perry, S., & Manson, J. H. (2003). Traditions in monkeys. *Evolutionary Anthropology: Issues, News, and Reviews*, *12*(2), 71–81. https://doi.org/10.1002/evan.10105

Perwitasari-Farajallah, D., Eka, A., Nurul, Q. I., Andre, P., Apri, A. D., & O, W. J. (2022). Pregnant and lactating *Macaca nigra*: Behavior and food selection. *BIOTROPIA*, *29*(2), Article 2. https://doi.org/10.11598/btb.2022.29.2.1687

Peterson, J. V., Fuentes, A., & Wandia, I. N. (2022). Cohort dominance rank and "robbing and bartering" among subadult male long-tailed macaques at Uluwatu, Bali. *Scientific Reports*, *12*(1), 7971. https://doi.org/10.1038/s41598-022-11776-7

Piet, G. J., Tamis, J. E., Volwater, J., de Vries, P., van der Wal, J. T., & Jongbloed, R. H. (2021). A roadmap towards quantitative cumulative impact assessments: Every step of the way. *Science of The Total Environment*, *784*, 146847. https://doi.org/10.1016/j.scitotenv.2021.146847

Pilgrim, G. E. (1915). *New Siwalik Primates and their bearing on the question of the evolution of man and the Anthropoidea*. [Data set]. Records of the Geological Survey of India.

Port, M., Hildenbrandt, H., Pen, I., Schülke, O., Ostner, J., & Weissing, F. J. (2020). The evolution of social philopatry in female primates. *American Journal of Physical Anthropology*. https://doi.org/10.1002/ajpa.24123

Portman, M. V. (1896). *A history of our relations with the Andamanese* (Vol. 1). Govt. of India Central Printing Office.

Porwal, M. C., Padalia, H., & Roy, P. S. (2012). Impact of tsunami on the forest and biodiversity richness in Nicobar Islands (Andaman and Nicobar Islands), India. *Biodiversity and Conservation*, *21*(5), 1267–1287. https://doi.org/10.1007/s10531-011-0214-x

Prakash, P. V. (2024). Archaic and lLiving traditions: Ethnoarchaeology of Andaman and Nicobar Archipelago. *The Oriental Anthropologist: A Bi-Annual International Journal of the Science of Man*, *24*(2), 187–206. https://doi.org/10.1177/0972558X241256795

Prakash, I. (1958). The breeding season of the Rhesus Monkey *Macaca Mulatta*, (Zimmerman) in Rajasthan. *The Journal of the Bombay Natural History Society*, *55*, 154.

Praveen Karanth, K. (2010). Molecular systematics and conservation of the langurs and leaf monkeys of South Asia. *Journal of Genetics*, *89*(4), Article 4. https://doi.org/10.1007/s12041-010-0057-3

Press Trust of India. (2023). *Mud Volcano at Baratang Erupts, Andaman and Nicobar: India's only Mud Volcano Erupts after 20 years In Andamans*. https://www.ndtv.com/india-news/mud-volcano-at-baratang-erupts-andaman-and-nicobar-indias-only-mud-volcano-erupts-after-20-years-in-andamans-9388583

Price, J. A. (1973). The superorganic fringe: Protoculture, indioculture, and material culture. *Ethos*, *1*(2), 201–218.

Prinz, J. F., & Lucas, P. W. (2000). Saliva tannin interactions. *Journal of Oral Rehabilitation*, *27*(11), 991–994. https://doi.org/10.1111/j.1365-2842.2000.00578.x

Priyadarshini, P., & Abhilash, P. C. (2019). Promoting tribal communities and indigenous knowledge as potential solutions for the sustainable development of India. *Environmental Development*, *32*, 100459. https://doi.org/10.1016/j.envdev.2019.100459

Proffitt, T., Luncz, L. V., Falótico, T., Ottoni, E. B., de la Torre, I., & Haslam, M. (2016). Wild monkeys flake stone tools. *Nature*, *539*(7627), 85–88. https://doi.org/10.1038/nature20112

Proffitt, T., Pacome, S. S., Reeves, J. S., Wittig, R. M., & Luncz, L. V. (2024). The archaeological visibility of chimpanzee (*Pan troglodytes*) nut-cracking. *Journal of Human Evolution*, *195*, 103582. https://doi.org/10.1016/j.jhevol.2024.103582

Proffitt, T., Reeves, Jonathan. S., Pacome, S. S., & Luncz, Lydia. V. (2022). Identifying functional and regional differences in chimpanzee stone tool technology. *Royal Society Open Science*, *9*(9), 220826. https://doi.org/10.1098/rsos.220826

Pyritz, L. W., Fichtel, C., Huchard, E., & Kappeler, P. M. (2013). Determinants and outcomes of decision-making, group coordination and social interactions during a foraging experiment in a wild primate. *PLoS One*, *8*(1), e53144. https://doi.org/10.1371/journal.pone.0053144

Rajeshkumar, S., Raghunathan, C., Chandra, K., & Venkataraman, K. (2014). Daily activity budget of Nicobar long-tailed macaque (*Macaca fascicularis umbrosa*) in Great Nicobar Island, India. *Journal of Research in Biology*, *4*(4), 1338–1347.

Ram, S., Venkatachalam, S., & Sinha, A. (2003). Changing social strategies of wild female bonnet macaques during natural foraging and on provisioning. *Current Science*, *84*(6), Article 6.

Ramakrishna, I. (2022, October 20). *Massive infrastructure project proposed by Central government threatens Great Nicobar Island*. Frontline. https://frontline.thehindu.com/environment/massive-infrastructure-project-proposed-by-central-government-threatens-great-nicobar-island/article65991977.ece

Ramakrishna, I., Mishra, P. S., Mazumder, J., & Pal, A. (2022). Macaca fascicularis ssp. Umbrosa*: The IUCN Red List of Threatened Species 2022: E.T39791A215086130* [Data set]. International Union for Conservation of Nature. https://doi.org/10.2305/IUCN.UK.2022-1.RLTS.T39791A215086130.en

Rapaport, L. G., & Brown, G. R. (2008). Social influences on foraging behavior in young nonhuman primates: Learning what, where, and how to eat. *Evolutionary Anthropology: Issues, News, and Reviews*, *17*(4), 189–201. https://doi.org/10.1002/evan.20180

Rathke, E.-M., & Fischer, J. (2021). Social aging in male and female Barbary macaques. *American Journal of Primatology*, *83*(11), e23272. https://doi.org/10.1002/ajp.23272

Rathke, E.-M., Mundry, R., & Fischer, J. (2022). Older Barbary macaques show limited capacity for self-regulation to avoid hazardous social interactions. *Communications Biology*, *5*(1), 1087. https://doi.org/10.1038/s42003-022-04012-5

Reeves, J. S., Proffitt, T., Malaivijitnond, S., & Luncz, L. V. (2023). Emergent technological variation in archaeological landscapes: A primate perspective. *Journal of The Royal Society Interface*, *20*(203), 20230118. https://doi.org/10.1098/rsif.2023.0118

Reeves, J. S., Proffitt, T., Pacome, S. S., & Luncz, L. V. (2024). Searching for the earliest archaeological record: Insights from chimpanzee material landscapes. *Journal of The Royal Society Interface*, *21*(217), 20240101. https://doi.org/10.1098/rsif.2024.0101

Reeves, J. S., Tan, A., Malaivijitnond, S., & Luncz, L. V. (2023). Simulation and social network analysis provide insight into the acquisition of tool behaviour in hybrid macaques. *Proceedings of the Royal Society B: Biological Sciences*, *290*(1995), 20222276. https://doi.org/10.1098/rspb.2022.2276

Reinegger, R. D., Oleksy, R. Z., Gazagne, E., & Jones, G. (2023). Foraging strategies of invasive macaca fascicularis may promote plant invasion in Mauritius. *International Journal of Primatology*, *44*(1), 140–170. https://doi.org/10.1007/s10764-022-00324-9

Ren, S., Liu, S., Sun, W., Gao, L., Ren, L., Liu, J., Zhang, W., Xia, D., Sun, B., Li, J., & Wang, X. (2024). Consistent individual differences drive collective movements in a Tibetan Macaque Group. *Animals*, *14*(10), 1476. https://doi.org/10.3390/ani14101476

Reyes, E., & Klug, H. (2021). Parental investment and sexual selection. In T. K. Shackelford & V. A. Weekes-Shackelford (Eds.), *Encyclopedia of evolutionary psychological science* (pp. 5711–5714). Springer International Publishing. https://doi.org/10.1007/978-3-319-19650-3_1691

Rifaie, F., Sulistyadi, E., Fitriana, Y. S., Inayah, N., Maharadatunkamsi, M., Prameswari, W., & Rusmadipraja, I. A. (2024). Spatial patterns of human–long-tailed macaque (*Macaca fascicularis*) conflicts in Java Island: A comparison of two secondary data sources. *Journal of Asia-Pacific Biodiversity*, *17*(4), 653–662. https://doi.org/10.1016/j.japb.2024.05.008

Riley, C. M., Koenig, B. L., & Gumert, M. D. (2015). Observation of a fatal dog attack on a juvenile long-tailed macaque in a human-modified environment.*Nature in Singapore*, *8*, 57–63.

Rizvi, S. N. (1990). *The Shompen: A vanishing tribe of the Great Nicobar Island*. Seagull Books.

Roberts, P., Delson, E., Miracle, P., Ditchfield, P., Roberts, R. G., Jacobs, Z., Blinkhorn, J., Ciochon, R. L., Fleagle, J. G., Frost, S. R., Gilbert, C. C., Gunnell, G. F., Harrison, T., Korisettar, R., & Petraglia, M. D. (2014). Continuity of mammalian fauna over the last 200,000 y in the Indian subcontinent. *Proceedings of the National Academy of Sciences*, *111*(16), 5848–5853. https://doi.org/10.1073/pnas.1323465111

Rodger, W. A., & Panwar, H. S. (1988). Planning a wildlife protected area network in India. Wildlife Institute of India.

Romanski, L. M. (2013). Integration of faces and vocalizations in ventral prefrontal cortex: Implications for the evolution of audiovisual speech. In G. F. Striedter, J. C. Avise, & F. J. Ayala (Eds.) *In the light of evolution: Volume VI: Brain and behavior*. National Academies Press (US). https://www.ncbi.nlm.nih.gov/books/NBK207155/

Rosien, J. E., Fannin, L. D., Yeakel, J. D., Malaivijitnond, S., Dominy, N. J., & Tan, A. (2025). Food-washing monkeys recognize the law of diminishing returns. *eLife*, *13*, RP98520. https://doi.org/10.7554/eLife.98520

Rothman, J. M., Dusinberre, K., & Pell, A. N. (2009). Condensed tannins in the diets of primates: A matter of methods? *American Journal of Primatology*, *71*(1), 70–76. https://doi.org/10.1002/ajp.20623

Roy, D. S., Krishnan, P., George, G., Kaliyamoorthy, M., & Bharathi, M. P. G. (2009). *Mangroves of Andaman and Nicobar Islands*. Indian Council of Agricultural Research.

Sahay, V. S. (2021). *Experiencing anthropology in the Nicobar Archipelago* (First published). Routledge.

Samuni, L., Preis, A., Mielke, A., Deschner, T., Wittig, R. M., & Crockford, C. (2018). Social bonds facilitate cooperative resource sharing in wild chimpanzees. *Proceedings of the Royal Society B: Biological Sciences*, *285*(1888), 20181643. https://doi.org/10.1098/rspb.2018.1643

Sanjib. (2023, April 23). Editorial: Water crisis: Time to take concrete steps to solve this issue for good. *ANDAMAN SHEEKHA*. https://www.andamansheekha.com/114488/

Sanz, C. M., Call, J., & Boesch, C. (Eds.). (2013). *Tool use in animals: Cognition and ecology*. Cambridge University Press.

Sanz, C., & Morgan, D. (2007). Chimpanzee tool technology in the Goualougo Triangle, Republic of Congo. *Journal of Human Evolution*, *52*(4), 420–433. https://doi.org/10.1016/j.jhevol.2006.11.001

Schaub, H. (1996). Testing kin altruism in long-tailed macaques (*Macaca fascicularis*) in a food-sharing experiment. *International Journal of Primatology*, *17*(3), 445–467. https://doi.org/10.1007/BF02736631

Schino, G. (2007). Grooming and agonistic support: A meta-analysis of primate reciprocal altruism. *Behavioral Ecology*, *18*(1), 115–120. https://doi.org/10.1093/beheco/arl045

Schino, G., & Aureli, F. (2008). Grooming reciprocation among female primates: A meta-analysis. *Biology Letters*, *4*(1), 9–11. https://doi.org/10.1098/rsbl.2007.0506

Schülke, O., Bhagavatula, J., Vigilant, L., & Ostner, J. (2010). Social bonds enhance reproductive success in Male Macaques. *Current Biology*, *20*(24), 2207–2210. https://doi.org/10.1016/j.cub.2010.10.058

Schülke, O., Ostner, J., & Berghänel, A. (2019). Prenatal maternal stress effects on the development of primate social behavior. *Behavioral Ecology and Sociobiology*, *73*(9). https://doi.org/10.1007/s00265-019-2729-1

Schuster, R. C., Wachter, K., Hussain, F., & Gartin, M. L. (2024). Gendered effects of climate change and health inequities among forcibly displaced populations: Displaced Rohingya women foster resilience through technology. *The Journal of Climate Change and Health*, *18*, 100303. https://doi.org/10.1016/j.joclim.2024.100303

Sekhsaria, P. (2024a). *GNI Timeline-May-2024 Revised-and-updated*. https://kalpavriksh.org/wp-content/uploads/2024/06/GNI-Timeline-May-2024

_Revised-and-updated.pdf?srsltid=AfmBOop5ZaNRE3AEhZCtPC2Ks2KOPLAHLxuphErYK8SXxvy2G4M-YD6E

Sekhsaria, P. (2024b). *The great Nicobar Betrayal*. THG Publishing Private Limited.

Sekhsaria, P., & Pandya, V. (Eds.). (2010). *The Jarawa Tribal Reserve Dossier: Cultural & biological diversities in the Andaman Islands*. http://unesdoc.unesco.org/images/0018/001876/187690e.pdf

Sen, S. (2010). *Savagery and colonialism in the Indian Ocean: Power, pleasure and the Andaman islanders*. Routledge.

Sen, U. (2017). Developing Terra Nullius: Colonialism, nationalism, and indigeneity in the Andaman Islands. *Comparative Studies in Society and History*, *59*(4). https://doi.org/10.1017/S0010417517000330

Sen, U. (Ed.). (2018). Harnessed to national development: Settlers, producers and agents of hinduisation. In *Citizen refugee: Forging the Indian Nation after partition* (pp. 70–112). Cambridge University Press. https://doi.org/10.1017/9781108348553.004

Sha, J. C. M., & Hanya, G. (2013a). Diet, activity, habitat use, and ranging of two neighboring groups of food-enhanced long-tailed macaques (*Macaca fascicularis*). *American Journal of Primatology*, *75*(6), 581–592. https://doi.org/10.1002/ajp.22137

Sha, J. C. M., & Hanya, G. (2013b). Temporal food resource correlates to the behavior and ecology of food-enhanced long-tailed macaques (*Macaca fascicularis*). *Mammal Study*, *38*(3), 163–175. https://doi.org/10.3106/041.038.0305

Shanker, K., Choudhury, B. C., & Aggarwal, R. K. (2011). *Conservation genetics of marine turtles on the mainland coast of India and offshore island*. WII Grant-in-Aid Project Report.

Sharief, M. U., & Panda, S. P. (2021). Folklore artefacts of aboriginal Shompen tribe in Great Nicobar Island, India. *Genetic Resources and Crop Evolution*, *68*(3), 835–847. https://doi.org/10.1007/s10722-020-01027-x

Sharma, D. (2020). *Tourism in the Andaman Islands: An assessment of challenges through two case studies*. Dakshin Foundation. https://www.dakshin.org/tourism-in-the-andaman-islands-an-assessment-of-challenges-through-two-case-studies/

Sharma, K., Kumar, V., Kaur, J., Tanwar, B., Goyal, A., Sharma, R., Gat, Y., & Kumar, A. (2021). Health effects, sources, utilization and safety of tannins: A critical review. *Toxin Reviews*, *40*(4), 432–444. https://doi.org/10.1080/15569543.2019.1662813

Sharma, P. (2024, July). Leatherback turtles: Impact of Airport on Leatherback Turtles' Nesting Site in Nicobar | Dehradun News-Times of India. *Times of India*. https://timesofindia.indiatimes.com/city/dehradun/impact-of-airport-on-leatherback-turtles-nesting-site-in-nicobar/articleshow/112121501.cms

Shashi, S. S. (Ed.). (1994). *Encyclopaedia of Indian Tribes*. Anmol Publ.

Shimada, M., & Sueur, C. (2018). Social play among juvenile wild Japanese macaques (*Macaca fuscata*) strengthens their social bonds. *American Journal of Primatology*, *80*(1), e22728. https://doi.org/10.1002/ajp.22728

Shumaker, R. W., Walkup, K. R., Beck, B. B., & Burghardt, G. M. (2011). *Animal tool behavior the use and manufacture of tools by animals*. The Johns Hopkins University Press.

Silk, J. B. (2007). Social components of fitness in primate groups. *Science*, *317*(5843), 1347–1351. https://doi.org/10.1126/science.1140734

Singh, M., Krishna, B. A., & Singh, M. (2006). Dominance hierarchy and social grooming in female lion- tailed macaques (*Macaca silenus*) in the Western Ghats, India. *Journal of Biosciences*, *31*(3), Article 3. https://doi.org/10.1007/BF02704110

Singh, R. K., & Sureja, A. K. (2008). Indigenous knowledge and sustainable agricultural resources management under rainfed agro-ecosystem. *Indian Journal of Traditional Knowledge*, *7*(4), 642–654

Singh, S. C., & Moeremans, R. (2017). Chapter 13 Anatomy of the Andaman–Nicobar subduction system from seismic reflection data. *Geological Society, London, Memoirs*, *47*(1), 193–204. https://doi.org/10.1144/M47.13

Sinha, A. (1997). Complex tool manufacture by a wild bonnet macaque, *Macaca radiata. Folia Primatologica*, *68*(1), 23–25. https://doi.org/10.1159/000157227

Sinha, A. (1998). Knowledge acquired and decisions made: Triadic interactions during allogrooming in wild bonnet macaques, Macaca radiata. *Philosophical Transactions of the Royal Society of London. Series B: Biological Sciences*, *353*(1368), 619–631. https://doi.org/10.1098/rstb.1998.0230

Siracusa, E. R., Negron-Del ValleJ. E., , Phillips, D., Platt, M. L., Higham, J. P., Snyder-Mackler, N., & Brent, L. J. N. (2022). Within-individual changes reveal increasing social selectivity with age in rhesus macaques. *Proceedings of the National Academy of Sciences*, *119*(49), e2209180119. https://doi.org/10.1073/pnas.2209180119

Sivakumar, K. (2010). Impact of the tsunami (December, 2004) on the long tailed macaque of Nicobar Islands, India. *Hystrix, the Italian Journal of Mammalogy*, *21*(1), Article 1. https://doi.org/10.4404/hystrix-21.1-4484

Slocombe, S. (2000). Conservation through cultural survival: Indigenous peoples and protected areas. *Mountain Research and Development*, *20*(2), 200–201. https://doi.org/10.1659/0276-4741(2000)020%255B0200:CTCSIP%255D2.0.CO;2

Sommer, V., & Mendoza-Granados, D. (2010). Play as indicator of habitat quality: A field study of langur monkeys (*Presbytis entellus*). *Ethology*, *99*(3), 177–192. https://doi.org/10.1111/j.1439-0310.1995.tb00893.x

Sonakia, A., & De Lumley, H. (2006). Narmada Homo erectus – A possible ancestor of the modern Indian. *Comptes Rendus Palevol*, *5*(1–2), 353–357. https://doi.org/10.1016/j.crpv.2005.10.002

Sonamuthu, K. (2019). Energy crisis in Andaman and Nicobar Islands. *International Advanced Research Journal in Science, Engineering and Technology*, *6*(10), 1–7.

Sterck, E. H. M., & Steenbeek, R. (1997). Female dominance relationships and food competition in the sympatric Thomas Langur and long-tailed macaque. *Behaviour*, *134*(9/10), 749–774.

Sterck, E. H. M., Watts, D. P., & Van Schaik, C. P. (1997). The evolution of female social relationships in nonhuman primates. *Behavioral Ecology and Sociobiology*, *41*(5), 291–309. https://doi.org/10.1007/s002650050390

Stewart, A.-M. E., Gordon, C. H., Wich, S. A., Schroor, P., & Meijaard, E. (2008). Fishing in *Macaca fascicularis*: A rarely observed innovative behavior. *International Journal of Primatology*, *29*(2), 543–548. https://doi.org/10.1007/s10764-007-9176-y

Strier, K. B. (2021). *Primate behavioral ecology* (6th ed). Taylor & Francis Group.

Stumpf, R. M., Martinez-Mota, R., Milich, K. M., Righini, N., & Shattuck, M. R. (2011). Sexual conflict in primates. *Evolutionary Anthropology: Issues, News, and Reviews*, *20*(2), 62–75. https://doi.org/10.1002/evan.20297

Sueur, C., Petit, O., De Marco, A., Jacobs, A. T., Watanabe, K., & Thierry, B. (2011). A comparative network analysis of social style in macaques. *Animal Behaviour*, *82*(4), 845–852. https://doi.org/10.1016/j.anbehav.2011.07.020

Sugiyama, Y. (1994). Age-specific birth rate and lifetime reproductive success of chimpanzees at Bossou, Guinea. *American Journal of Primatology*, *32*(4), 311–318. https://doi.org/10.1002/ajp.1350320408

Sullivan, L. (2019). Conservation in context: Toward a systems framing of decentralized governance and public participation in wildlife management. *Review of Policy Research*, *36*(2), 242–261. https://doi.org/10.1111/ropr.12326

Swaminathan, A. (2021). *Leatherback Sea Turtles: Giants of Andaman and Nicobar Islands | Roundglass Sustain | Roundglass | Sustain*. https://roundglasssustain.com/conservation/protecting-leatherback-sea-turtles

Swedell, L. (2012). Primate sociality and social systems. *Nature Education Knowledge*, *3*(10), 84

Tacconi, L. (2007). *Illegal logging: Law enforcement, livelihoods and the timber trade*. Earthscan. http://www.123library.org/book_details/?id=85592

Tan, A., Tan, S. H., Vyas, D., Malaivijitnond, S., & Gumert, M. D. (2015). There is more than one way to crack an oyster: Identifying variation in burmese long-tailed macaque (*Macaca fascicularis aurea*) stone-tool use. *PLoS One*, *10*(5), e0124733. https://doi.org/10.1371/journal.pone.0124733

Tan, A. W. Y. (2016). *Behavioral processes and social influences on the development of stone-tool use in long-tailed macaques* [Nanyang Technological University]. https://doi.org/10.32657/10356/69347

Tan, A. W. Y. (2017). From play to proficiency: The ontogeny of stone-tool use in coastal-foraging long-tailed macaques (*Macaca fascicularis*) from a comparative perception-action perspective. *Journal of Comparative Psychology*, *131*(2), 89–114. https://doi.org/10.1037/com0000068

Tan, A. W. Y., Hemelrijk, C. K., Malaivijitnond, S., & Gumert, M. D. (2018). Young macaques (*Macaca fascicularis*) preferentially bias attention towards closer, older, and better tool users. *Animal Cognition*, *21*(4), 551–563. https://doi.org/10.1007/s10071-018-1188-9

Tennie, C., Jensen, K., & Call, J. (2016). The nature of prosociality in chimpanzees. *Nature Communications*, *7*, 13915. https://doi.org/10.1038/ncomms13915

Thierry, B. (2007). Unity in diversity: Lessons from macaque societies. *Evolutionary Anthropology: Issues, News, and Reviews*, *16*(6), 224–238. https://doi.org/10.1002/evan.20147

Thierry, B., Aureli, F., Nunn, C. L., Petit, O., Abegg, C., & de Waal, F. B. M. (2008). A comparative study of conflict resolution in macaques: Insights into the nature of trait covariation. *Animal Behaviour*, *75*(3), 847–860. https://doi.org/10.1016/j.anbehav.2007.07.006

Thiollay, J. (1997). Distribution and abundance patterns of bird community and raptor populations in the Andaman archipelago. *Ecography*, *20*(1), 67–82. https://doi.org/10.1111/j.1600-0587.1997.tb00348.x

Thiollay, J. (1998). Distribution patterns and insular biogeography of South Asian raptor communities. *Journal of Biogeography*, *25*(1), 52–72. https://doi.org/10.1046/j.1365-2699.1998.251164.x

Thirumurugan, V., Singh, A. R., & Prabakaran, N. (2022). First report on the occurrence of Avicennia marina (Forssk.) Vierh. (Acanthaceae) in the Nicobar archipelago. *Ocean and Coastal Research*, *70*, e22013. https://doi.org/10.1590/2675-2824070.21077vt

Trevathan, W. (2015). Primate pelvic anatomy and implications for birth. *Philosophical Transactions of the Royal Society B: Biological Sciences*, *370*(1663), 20140065. https://doi.org/10.1098/rstb.2014.0065

Trivers, R. (1972). Parental investment and sexual selection. In B. Campbell (Ed.), *Sexual selection and the descent of man* (pp. 136–179).

Tsuchida, J., Yoshida, T., Sankai, T., & Yasutomi, Y. (2008). Maternal behavior of laboratory-born, individually reared long-tailed macaques (*Macaca fascicularis*). *Journal of the American Association for Laboratory Animal Science: JAALAS*, *47*(5), 29–34.

Uehara, M., Tabata, E., Okuda, M., Maruyama, Y., Matoska, V., Bauer, P. O., & Oyama, F. (2021). Robust chitinolytic activity of crab-eating monkey (*Macaca fascicularis*) acidic chitinase under a broad pH and temperature range. *Scientific Reports*, *11*(1), 15470. https://doi.org/10.1038/s41598-021-95010-w

Urbani, B., Youlatos, D., & Antczak, A. T. (Eds.). (2022). *World archaeoprimatology: Interconnections of humans and nonhuman primates in the past* (1st ed.). Cambridge University Press. https://doi.org/10.1017/9781108766500

V., K., Priya, P. R., K., T., Mazumder, B., Gayatri, P., & Neha, J. (2023). Assessment of water demand and potential water sources to face future water scarcity of hilly regions. *Journal of Water Supply: Research and Technology-Aqua*, jws2023212. https://doi.org/10.2166/aqua.2023.212

Vale, G. L., Carr, K., Dean, L. G., & Kendal, R. L. (2017). The cultural capacity of human and nonhuman primates: Social learning, innovation, and cumulative cultural evolution. In J. H. Kaas (Ed.), *Evolution of nervous systems* (pp. 475–508). Elsevier. https://doi.org/10.1016/B978-0-12-804042-3.00095-6

Van Buskirk, J. (2012). *Behavioural plasticity and environmental change*. https://doi.org/10.5167/UZH-64497

van Noordwijk, M. A. V., & van Schaik, C. P. V. (2001). Career moves: Transfer and rank challenge decisions by male long-tailed macaques. *Behaviour*, *138*(3), 359–395.

van Noordwijk, M. A., & van Schaik, C. P. (1985). Male migration and rank acquisition in wild long-tailed macaques (*Macaca fascicularis*). *Animal Behaviour*, *33*(3), 849–861. https://doi.org/10.1016/S0003-3472(85)80019-1

Van Noordwijk, M. A., & Van Schaik, C. P. (1986). The hidden costs of sociality: Intra-group variation in feeding strategies in sumatran long-tailed macaques (*Macaca fascicularis*). *Behaviour*, *99*(3–4), 296–314. https://doi.org/10.1163/156853986X00595

van Noordwijk, M. A., & van Schaik, C. P. (1987). Competition among female long-tailed macaques, Macaca fascicularis. *Animal Behaviour*, *35*(2), 577–589. https://doi.org/10.1016/S0003-3472(87)80284-1

van Noordwijk, M. A., & van Schaik, C. P. (1999). The effects of dominance rank and group size on female lifetime reproductive success in wild long-tailed macaques, *Macaca fascicularis*. *Primates*, *40*(1), 105–130. https://doi.org/10.1007/BF02557705

van Schaik, C. P., Ancrenaz, M., Borgen, G., Galdikas, B., Knott, C. D., Singleton, I., Suzuki, A., Utami, S. S., & Merrill, M. (2003). Orangutan cultures and the evolution of material culture. *Science*, *299*(5603), 102–105. https://doi.org/10.1126/science.1078004

van Schaik, C. P., & Brockman, D. K. (Eds.). (2005). *Seasonality in primates: Studies of living and extinct human and non-human primates*. Cambridge University Press.

van Schaik, C. P., & Burkart, J. M. (2011). Social learning and evolution: The cultural intelligence hypothesis. *Philosophical Transactions of the Royal Society*

B: Biological Sciences, *366*(1567), 1008–1016. https://doi.org/10.1098/rstb.2010.0304
Van Schaik, C. P., Pandit, S. A., & Vogel, E. R. (2004). A model for within-group coalitionary aggression among males. *Behavioral Ecology and Sociobiology*, *57*(2), 101–109. https://doi.org/10.1007/s00265-004-0818-1
van Schaik, C. P., & Pradhan, G. R. (2003). A model for tool-use traditions in primates: Implications for the coevolution of culture and cognition. *Journal of Human Evolution*, *44*(6), 645–664. https://doi.org/10.1016/S0047-2484(03)00041-1
Vauclair, J., & Anderson, J. (1994). Object manipulation, tool use, and the social context in human and non-human primates. *Techniques and Culture*, *23–24*, 121–136. https://doi.org/10.4000%2520/%2520tc.556
Veenema, H. C., Spruijt, B. M., Gispen, W. H., & Van Hooff, J. A. R. A. M. (1997). Aging, dominance history, and social behavior in Java-Monkeys (Macaca fascicularis). *Neurobiology of Aging*, *18*(5), 509–515. https://doi.org/10.1016/S0197-4580(97)00107-3
Veenema, H. C., van Hooff, J. A. R. A. M., Gispen, W. H., & Spruijt, B. M. (2001). Increased rigidity with age in social behavior of Java-monkeys (*Macaca fascicularis*). *Neurobiology of Aging*, *22*(2), 273–281. https://doi.org/10.1016/S0197-4580(00)00204-9
Velankar, A. D., Kumara, H. N., Pal, A., Mishra, P. S., & Singh, M. (2016). Population recovery of nicobar long-tailed macaque *Macaca fascicularis umbrosus* following a Tsunami in the Nicobar Islands, India. *PLoS One*, *11*(2), e0148205. https://doi.org/10.1371/journal.pone.0148205
Velankar, A. D., Kumara, H. N., Pal, A., Mishra, P. S., & Singh, M. (2023). A tough nut to crack? Proficiency in coconut processing of the Nicobar long-tailed macaque Macaca fascicularis umbrosus. *Journal of Biosciences*, *48*(3), 28. https://doi.org/10.1007/s12038-023-00352-4
Velankar, A. D., Kumara, H. N., Sheethal, V. R., Pal, A., Mishra, P. S., & Singh, M. (2024). Activity budget and habitat use in the Nicobar long-tailed macaque Macaca fascicularis umbrosus. *Journal of Wildlife Science*, *1*(2), 87–95. https://doi.org/10.63033/JWLS.RUPT1872
Verma, S., Das, A., Sachin, R., & Pal, T. (2023). Preliminary appraisal of deformation structures from Great Nicobar Island, Andaman-Sumatra arc system. *Results in Earth Sciences*, *1*, 100007. https://doi.org/10.1016/j.rines.2023.100007
Viaro, R., Bernardi, D., Mazzoni, L., & Fadiga, L. (2023). Interaction-mediated tool use differently enhances physical and social cognition in macaques (Macaca fascicularis). *International Journal of Primatology*, *44*(5), 928–959. https://doi.org/10.1007/s10764-023-00371-w
Vos, D. R., Karssemeijer, G. J., & van Hooff, J. A. R. A. M. (1992). Ecological constraints on the behaviour of mother long-tailed macaques (Macaca fascicularis). *Behavioral Ecology and Sociobiology*, *31*(6), 385–391. https://doi.org/10.1007/BF00170605
Waal, F. B. M. de, & Ferrari, P. F. (Eds.). (2012). *The primate mind: Built to connect with other minds*. Harvard University Press.
Walther, G.-R., Post, E., Convey, P., Menzel, A., Parmesan, C., Beebee, T. J. C., Fromentin, J.-M., Hoegh-Guldberg, O., & Bairlein, F. (2002). Ecological responses to recent climate change. *Nature*, *416*(6879), 389–395. https://doi.org/10.1038/416389a
Wang, W., Ma, L., Versteegh, M. A., Wu, H., & Komdeur, J. (2021). Parental care system and brood size drive sex difference in reproductive allocation: An

experimental study on burying beetles. *Frontiers in Ecology and Evolution*, *9*, 739396. https://doi.org/10.3389/fevo.2021.739396
Watanabe, K., Urasopon, N., & Malaivijitnond, S. (2007). Long-tailed macaques use human hair as dental floss. *American Journal of Primatology*, *69*(8), 940–944. https://doi.org/10.1002/ajp.20403
Wcislo, W. T. (1989). Behavioral environments and evolutionary change. *Annual Review of Ecology and Systematics*, *20*, 137–169.
Welker, C. (1992). Long-term studies on the social behavior of the Crab-eating monkey, *Macaca fasciularis*. In P. K. Seth (Ed.), *Perspectives in Primate Biology* (Vol. 4, pp. 17–26). Today & Tomorro's Printers and Publishers.
Westberg, G. (2023). Desiring the indigenous: Affective commodification of the Sámi. *Social Semiotics*, *33*(4), 769–786. https://doi.org/10.1080/10350330.2021.1942824
Wheatley, B. P. (1989). Diet of Balinese temple monkeys, *Macaca fascicularis*. *Kyoto Univ Overseas Res Rep Stud Asian Non-Human Primates*, *7*, 62–75.
Wheatley, B. P. (1999). *The sacred monkeys of Bali*. Waveland Press.
Whitehouse, J., Clark, P. R., Robinson, R. L., Rees, K., O'Callaghan, O., Kimock, C. M., Witham, C. L., & Waller, B. M. (2024). Facial expressivity in dominant macaques is linked to group cohesion. *Proceedings of the Royal Society B: Biological Sciences*, *291*(2027), 20240984. https://doi.org/10.1098/rspb.2024.0984
Whiten, A. (2013). Archaeology meets primate technology: Behavioural biology. *Nature*, *498*(7454), 303–305. https://doi.org/10.1038/498303a
Whiten, A. (2020). A unified account of culture should accommodate animal cultures. *Behavioral and Brain Sciences*, *43*, e118. https://doi.org/10.1017/S0140525X1900270X
Whiten, A., Goodall, J., McGrew, W. C., Nishida, T., Reynolds, V., Sugiyama, Y., Tutin, C. E. G., Wrangham, R. W., & Boesch, C. (1999). Cultures in chimpanzees. *Nature*, *399*(6737), 682–685. https://doi.org/10.1038/21415
Whiten, A., Horner, V., & de Waal, F. B. M. (2005). Conformity to cultural norms of tool use in chimpanzees. *Nature*, *437*(7059), 737–740. https://doi.org/10.1038/nature04047
Whiten, A., & Van De Waal, E. (2018). The pervasive role of social learning in primate lifetime development. *Behavioral Ecology and Sociobiology*, *72*(5), 80. https://doi.org/10.1007/s00265-018-2489-3
Wong, B. B. M., & Candolin, U. (2015). Behavioral responses to changing environments. *Behavioral Ecology*, *26*(3), 665–673. https://doi.org/10.1093/beheco/aru183
Wooddell, L. J., Kaburu, S. S. K., & Dettmer, A. M. (2019). Dominance rank predicts social network position across developmental stages in rhesus monkeys. *American Journal of Primatology*, e23024. https://doi.org/10.1002/ajp.23024
Wrangham, R. W. (1980). An ecological model of female-bonded primate groups. *Behaviour*, *75*(3–4), 262–300. https://doi.org/10.1163/156853980X00447
Wright, K. R., Mayhew, J. A., Sheeran, L. K., Funkhouser, J. A., Wagner, R. S., Sun, L.-X., & Li, J.-H. (2018). Playing it cool: Characterizing social play, bout termination, and candidate play signals of juvenile and infant Tibetan macaques (Macaca thibetana). *Zoological Research*, *39*(4), 272–283. https://doi.org/10.24272/j.issn.2095-8137.2018.048
Zhang, K., Karim, F., Jin, Z., Xiao, H., Yao, Y., Ni, Q., Li, B., Pu-Cuo, W., Huang, Z., & Xu, H. (2022). Diet and feeding behavior of a group of high-altitude rhesus macaques: High adaptation to food shortages and seasonal fluctuations. *Current Zoology*, zoac047. https://doi.org/10.1093/cz/zoac047

Zhang, T., Liu, S.-Q., Xia, Y.-N., Li, B.-W., Wang, X., & Li, J.-H. (2023). Aging-related behavioral patterns in tibetan macaques. *Biology*, *12*(10), 1325. https://doi.org/10.3390/biology12101325

Zhang, Y., Westaway, K. E., Haberle, S., Lubeek, J. K., Bailey, M., Ciochon, R., Morley, M. W., Roberts, P., Zhao, J., Duval, M., Dosseto, A., Pan, Y., Rule, S., Liao, W., Gully, G. A., Lucas, M., Mo, J., Yang, L., Cai, Y., … Joannes-Boyau, R. (2024). The demise of the giant ape Gigantopithecus blacki. *Nature*, *625*(7995), 535–539. https://doi.org/10.1038/s41586-023-06900-0

Zhuang, X., Yao, Y., & Li, J. (Justin). (2019). Sociocultural impacts of tourism on residents of world cultural heritage sites in China. *Sustainability*, *11*(3), Article 3. https://doi.org/10.3390/su11030840

Zipple, M. N., Altmann, J., Campos, F. A., Cords, M., Fedigan, L. M., Lawler, R. R., Lonsdorf, E. V., Perry, S., Pusey, A. E., Stoinski, T. S., Strier, K. B., & Alberts, S. C. (2021). Maternal death and offspring fitness in multiple wild primates. *Proceedings of the National Academy of Sciences*, *118*(1), e2015317118. https://doi.org/10.1073/pnas.2015317118

Index

For Product Safety Concerns and Information please contact our EU representative GPSR@taylorandfrancis.com
Taylor & Francis Verlag GmbH, Kaufingerstraße 24, 80331 München, Germany

www.ingramcontent.com/pod-product-compliance
Ingram Content Group UK Ltd.
Pitfield, Milton Keynes, MK11 3LW, UK
UKHW020134200926
475702UK00005B/750
* 9 7 8 1 0 4 1 0 1 4 0 1 0 *